STUDENT SOLUTIONS MANUAL

CALCULUS

from Graphical, Numerical, and Symbolic Points of View

Volume 2

Arnold Ostebee
Paul Zorn

St. Olaf College

Saunders College Publishing

Harcourt Brace College Publishers

Fort Worth Philadelphia San Diego New York Orlando Austin
San Antonio Toronto Montreal London Sydney Tokyo

Ostebee & Zorn: Student Solutions Manual to accompany *Calculus from Graphical, Numerical, and Symbolic Points of View*.

ISBN 0-03-017433-3

7 021 98765432

5.1   Areas and Integrals

1. The area of the entire rectangle shown is only 250, so $\int_1^2 g(x)\,dx < 250$. Furthermore, since $g(x) > 0$ for all x, $\int_1^2 g(x)\,dx > 0$. Finally, it is clear from the picture that $\int_{1.75}^2 g(x)\,dx > 12.5$ (the area of one dotted rectangle). Thus, the only possible choice is $\int_1^2 g(x)\,dx \approx 45$.

3. (a) Since $2 < f(x) < 5$ on $[1, 6]$, $10 < \int_1^6 f(x)\,dx < 25$. Thus, the best estimate for the value of the integral is 20. [NOTE: Be sure to take into account that the interval of integration is $[1, 6]$ and not $[0, 8]$!]

    (b) $A = 12$ and $B = 20$ since $3 \le f(x) \le 5$ if $3 \le x \le 7$. (Many other answers are possible.)

    (c) This approximation underestimates the exact value of the integral. (Count squares.)

    (d) $\frac{1}{2}\int_0^2 f(x)\,dx \approx \frac{11}{2}$

    (e) too large

5. (a) $\int_0^2 f(x)\,dx = 4$ since this definite integral is the area of a trapezoid with base 2 and heights 1 and 3.

    (b) $\int_2^5 f(x)\,dx = 9\pi/4$ since this definite integral is the area of a quarter-circle with radius 3.

    (c) $\int_0^5 f(x)\,dx = \int_0^2 f(x)\,dx + \int_2^5 f(x)\,dx = 4 + 9\pi/4$

    (d) $\int_5^9 f(x)\,dx = -4\pi$ since this definite integral is the area of a quarter-circle with radius 4 that lies below the x-axis.

    (e) $\int_5^5 f(x)\,dx = 0$

    (f) $\int_0^{15} f(x)\,dx = \int_0^2 f(x)\,dx + \int_2^5 f(x)\,dx + \int_5^9 f(x)\,dx + \int_9^{15} f(x)\,dx$

    $\qquad\qquad\quad = 4 + 9\pi/4 - 4\pi - 12 \approx -13.5$

    (g) $\int_0^{15} |f(x)|\,dx = \int_0^2 f(x)\,dx + \int_2^5 f(x)\,dx - \int_5^9 f(x)\,dx - \int_9^{15} f(x)\,dx$

    $\qquad\qquad\qquad = 4 + 9\pi/4 + 4\pi + 12 \approx 35.6$

    since $|f(x)| = -f(x)$ if $f(x) < 0$.

    (h) $\int_{15}^9 f(x)\,dx = -\int_9^{15} f(x)\,dx = 12$ since the second definite integral is the area of a triangle with base 6 and height 4 that lies below the x-axis.

    (i) $\int_{12}^{15} f(x)\,dx = -3$

    (j) $\int_9^{12} f(x)\,dx = -9$ since this definite integral is the area of a trapezoid with base 3 and heights 4 and 2 that lies below the x-axis.

7. Jack's answer is too big — the area of the entire rectangle shown is only $\pi/2$. Since $\cos^8 x \ge 0$ on the entire interval $[0, \pi/2]$, the value of the integral must be positive. This rules out Ed's answer. Furthermore, the value of the integral is approximately the area of the triangle with vertices at $(0, 0)$, $(0, 1)$, and $(1, 0)$. Since this triangle has area $1/2$, Lesley's answer is too big. Therefore, Joan's answer must be the correct answer.

    [NOTE: Since $\cos^8 x \le 1$ if $0 \le x \le 1/2$ and $\cos^8 x \le 2/5$ if $1/2 \le x \le 1$, $\int_0^1 \cos^8 x\,dx = \int_0^{1/2} \cos^8 x\,dx + \int_{1/2}^1 \cos^8 x\,dx \le 1/2 + 1/5 = 7/10$.]

9. (a) $h'(5) \approx -2.2$

    (b) $\frac{1}{10}\int_0^{10} h(x)\,dx \approx 17/10 = 1.7$

    (c) $\frac{1}{10}\big(h(10) - h(0)\big) = -4/10 = -0.4$

    (d) $\int_0^{10} h(x)\,dx \approx 17$

    (e) $\int_0^5 h(x)\,dx \approx 20$

    (f) $\int_6^{10} h(x)\,dx \approx -4.7$

11. (a) $\int_0^4 v(t)\,dt = 80$. At time $t = 4$ the car is 80 miles east of its starting point.

(b) The car's average (eastward) velocity is 20 mph $= \frac{1}{4} \int_0^4 v(t)\, dt$.

(c) Since $s(t) = |v(t)|$, $\int_0^4 s(t)\, dt = 100$. Over the 4 hours, the car travels a total distance of 100 miles.

(d) 100 miles/4 hours = 25 mph

13. (a) $\int_0^2 f(x)\, dx = 2$, $\int_1^4 f(x)\, dx = 3$, $\int_{-5}^{-1} f(x)\, dx = 4$, $\int_{-2}^3 f(x)\, dx = 5$

(b) $\int_0^2 f(x)\, dx = 2a$, $\int_1^4 f(x)\, dx = 3a$, $\int_{-5}^{-1} f(x)\, dx = 4a$, $\int_{-2}^3 f(x)\, dx = 5a$

(c) $\int_0^2 f(x)\, dx = -2a$, $\int_1^4 f(x)\, dx = -3a$, $\int_{-5}^{-1} f(x)\, dx = -4a$, $\int_{-2}^3 f(x)\, dx = -5a$

15. (a) $\int_{-3}^3 (x+2)\, dx = 12$ (c) $\int_{-3}^3 (|x|+2)\, dx = 21$

(b) $\int_{-3}^3 |x+2|\, dx = 13$ (d) $\int_{-3}^3 (2-|x|)\, dx = 3$

17. $\int_0^1 \sqrt{1-(x-1)^2}\, dx = \pi/4$

[NOTE: The integral is the area of a quarter-circle with radius 1 and center at $(1, 0)$.]

19. $\int_0^3 f(x)\, dx = 3 + 3\pi/4$

21. (a) $\int_1^4 f(x)\, dx = \int_1^2 f(x)\, dx + \int_2^4 f(x)\, dx = -1 + 7 = 6$

(b) $\int_0^4 3f(x)\, dx = 3 \left(\int_0^2 f(x)\, dx + \int_2^4 f(x)\, dx \right) = 3 \cdot 9 = 27.$

(c) $\int_0^1 f(x)\, dx = \int_0^2 f(x)\, dx - \int_1^2 f(x)\, dx = 2 - (-1) = 3.$

(d) $\int_0^1 f(x+1)\, dx = \int_1^2 f(x)\, dx = -1$

(e) $\int_0^1 \big(f(x)+1\big)\, dx = \int_0^1 f(x)\, dx + \int_0^1 dx = 3 + 1 = 4.$

(f) $\int_2^4 f(x-2)\, dx = \int_0^2 f(x)\, dx = 2.$

(g) $\int_2^4 \big(f(x)-2\big)\, dx = \int_2^4 f(x)\, dx - 2\int_2^4 dx = 7 - 2 \cdot 2 = 3.$

(h) $-1 = \int_1^2 f(x)\, dx < 0$ implies that $f(x) < 0$ over some (or all) of the interval $[1, 2]$.

(i) $6 = \int_2^4 3\, dx < \int_2^4 f(x)\, dx = 7$ implies that $f(x) > 3$ over some (or all) of the interval $[0, 2]$.

(j) One possibility: $f(x) = \begin{cases} -8(x-1)-1 & 0 \le x \le 1 \\ -1 & 1 < x \le 2 \\ 9(x-2)-1 & x > 2 \end{cases}$

23. (a) $\int_0^\pi \cos x\, dx = 0$ (c) $\int_{-2}^2 \left(7x^5 + 3\right) dx = 12$

(b) $\int_{\pi/2}^{3\pi/2} \sin x\, dx = 0$ (d) $\int_{-1}^1 \left(4x^3 - 2x\right) dx = 0$

25. $\int_1^3 \frac{1-x}{x^2}\, dx = \int_1^2 \frac{1-x}{x^2}\, dx + \int_2^3 \frac{1-x}{x^2}\, dx$. Since the integrand $((1-x)/x^2)$ is negative over the interval $[2, 3]$, $\int_2^3 \frac{1-x}{x^2}\, dx < 0$. This implies that $\int_1^3 \frac{1-x}{x^2}\, dx < \int_1^2 \frac{1-x}{x^2}\, dx$.

27. (a) $1 + \cos x \ge 0$ for all x.

(b) Since $\int_0^\pi \cos x\, dx = \int_\pi^{2\pi} \cos x\, dx = 0$, $\int_0^{2\pi} (1 + \cos x)\, dx = \int_0^{2\pi} dx = 2\pi > 0.$

29. Let $f(x) = e^x \sin x$. Then, $2.25 < e \sin 1 \le f(x) \le e^{3\pi/4} \dfrac{\sqrt{2}}{2} < 7.5$ if $1 \le x \le 3$. Therefore,

$4.5 = 2 \cdot 2.25 \le \int_1^3 f(x)\, dx \le 2 \cdot 7.5 = 15.$

31. Since $1/2 \le \sin x \le 1$ if $\pi/6 \le x \le \pi/2$, $\pi/6 = \frac{1}{2} \cdot (\pi/2 - \pi/6) \le \int_{\pi/6}^{\pi/2} \sin x \, dx \le 1 \cdot (\pi/2 - \pi/6) = \pi/3$.

33. (a) If $0 \le x \le \pi$, then $0 \le \sin x \le 1 \implies 1 \le e^{\sin x} \le e$. Thus, $\pi \le \int_0^\pi e^{\sin x} \, dx \le \pi e$.

   If $\pi \le x \le 2\pi$, $-1 \le \sin x \le 0 \implies 1/e \le e^{\sin x} \le 1$. Thus, $\pi/e \le \int_0^\pi e^{\sin x} \, dx \le \pi$.

   Since $\int_0^{2\pi} e^{\sin x} \, dx = \int_0^\pi e^{\sin x} \, dx + \int_\pi^{2\pi} e^{\sin x} \, dx$, the inequalities above lead to

   $\pi(1 + 1/e) \le \int_0^{2\pi} e^{\sin x} \, dx \le \pi(1 + e)$.

   (b) Since the integrand is periodic with period 2π, $25\pi(1 + 1/e) \le \displaystyle\int_0^{50\pi} e^{\sin x} \, dx \le 25\pi(1 + e)$.

35. Let $f(x) = \sin(e^x)$. Then, $0.4 < \sin(e) \le f(x) \le 1$ if $0 \le x \le 1$. Therefore, $0.4 \le \displaystyle\int_0^1 f(x) \, dx \le 1$.

37. If $0 \le x \le \pi$, $\cos 1 \le \cos(\sin x) \le 1$. Therefore,

$$\frac{\pi}{2} < \cos 1 \cdot (\pi - 0) \le \int_0^\pi \cos(\sin x) \, dx \le 1 \cdot (\pi - 0) = \pi.$$

39. (a) Let $f(x) = \sin x - 2x/\pi$. Then $f'(x) = \cos x - 2/\pi$, so f has a single stationary point (a local maximum) in the interior of the interval $[0, 1]$. Since $f(0) = 0$ and $f(1) > 0$, $2x/\pi \le \sin x$ if $0 \le x \le 1$.

   Let $g(x) = x - \sin x$. Then $g'(x) = 1 - \cos x \ge 0$ if $0 \le x \le 1$. Since $g(0) = 0$, $\sin x \le x$ if $0 \le x \le 1$.

   (b) Since $\displaystyle\int_0^{\pi/4} x \, dx = \pi^2/32$, the results in part (a) lead directly to the desired inequalities.

41. (a) Since $0 \le x^k f(x) \le f(x)$ if $0 \le x \le 1$ and $k \ge 0$ is an integer, $\int_0^1 x^k f(x) \, dx \le \int_0^1 f(x) \, dx$.

   (b) No. For example, let $f(x) = -1$. Then $-1 = \int_0^1 f(x) \, dx < \int_0^1 x f(x) \, dx = -1/2$.

43. (a) $\displaystyle\int_0^{\pi/2} \cos^2 x \, dx = \int_0^{\pi/2} \sin^2(x - \pi/2) \, dx = \int_{-\pi/2}^0 \sin^2 x \, dx = \int_0^{\pi/2} \sin^2 x \, dx$

   (b) $\displaystyle\int_0^{\pi/2} \sin^2 x \, dx + \int_0^{\pi/2} \cos^2 x \, dx = \int_0^{\pi/2} \left(\sin^2 x + \cos^2 x \right) dx = \int_0^{\pi/2} dx = \frac{\pi}{2}$

   (c) $\displaystyle\int_0^{\pi/2} \sin^2 x \, dx + \int_0^{\pi/2} \cos^2 x \, dx = 2 \int_0^{\pi/2} \sin^2 x \, dx = \frac{\pi}{2}$ so $\int_0^{\pi/2} \sin^2 x \, dx = \frac{\pi}{4}$.

45. Since f is an odd function, $\displaystyle\int_{-a}^0 f(x) \, dx = -\int_0^a f(x) \, dx$. Thus, $\int_{-a}^a f(x) \, dx = 0$.

47. The bounds on f imply that $-4 \le \displaystyle\int_1^3 f(x) \, dx \le 10$. Thus, $-2 \le \dfrac{\int_1^3 f(x) \, dx}{2} \le 5$.

49. $\int_{-3}^1 f(x) \, dx = 2 \cdot 4 = 8$ and $\int_{-3}^7 f(x) \, dx = 10 \cdot 5 = 50$, so $\int_1^7 f(x) \, dx = 42$. Therefore, the average value of f over the interval $[1, 7]$ is $42/6 = 7$.

51. Both integrals measure the area of the same region.

53. (a) No. Let $f(x) = 0$ and $g(x) = x$. Then $\int_{-1}^1 f(x) \, dx = \int_{-1}^1 g(x) \, dx = 0$ but $f(x) \ge g(x)$ if $-1 \le x \le 0$.

   (b) Yes. If $f(x) > g(x)$ for *every* x such that $a \le x \le b$, then $\int_a^b f(x) \, dx > \int_a^b g(x) \, dx$ would be true — a contradiction.

55. (b) $\displaystyle\int_1^2 f(x) \, dx = 6$

   $\displaystyle\int_1^3 f(x) \, dx = 14$

(c) $x = 2f^{-1}(x) + 3 \implies f^{-1}(x) = \frac{1}{2}(x - 3)$

(d) $\displaystyle\int_5^9 f^{-1}(x)\,dx = 8$

$\displaystyle\int_5^7 f^{-1}(x)\,dx = 3$

(e) $2f(2) - f(1) - \displaystyle\int_{f(1)}^{f(2)} f^{-1}(x)\,dx = 2 \cdot 7 - 1 \cdot 5 - \int_5^7 f^{-1}(x)\,dx = 14 - 5 - 3 = 6 = \int_1^2 f(x)\,dx$

$3f(3) - f(1) - \displaystyle\int_{f(1)}^{f(3)} f^{-1}(x)\,dx = 3 \cdot 9 - 1 \cdot 5 - \int_5^9 f^{-1}(x)\,dx = 27 - 5 - 8 = 14 = \int_1^3 f(x)\,dx$

57. $\displaystyle\int_0^a \sqrt{x}\,dx = a^{3/2} - \int_0^{\sqrt{a}} x^2\,dx = a^{3/2} - \frac{1}{3}a^{3/2} = \frac{2}{3}a^{3/2}.$

5.2 The Area Function

1. Suppose that $x < 0$. Then, $\int_x^0 t\,dt = -x^2/2$ and, therefore $\int_0^x t\,dt = x^2/2$.

3. (a) $F(x) = ax$; $G(x) = a(x-2)$; $H(x) = a(x+1)$. Yes.

   (b) $F(x) = bx^2/2$; $G(x) = b(x^2-4)/2$; $H(x) = b(x^2-1)/2$. Yes.

   (c) $F(x) = bx^2/2 + ax$; $G(x) = b(x^2-4)/2 + a(x-2)$; $H(x) = b(x^2-1)/2 + a(x+1)$. Yes.

5. $G(x) = \displaystyle\int_b^x f(t)\,dt = \int_a^x f(t)\,dt - \int_a^b f(t)\,dt = F(x) + C$ where $C = -\displaystyle\int_a^b f(t)\,dt$.

7. (a) $A_f(\pi) = 2$, $A_f(3\pi/2) = 1$, $A_f(2\pi) = 0$, $A_f(-\pi/2) = 1$, $A_f(-\pi) = 2$, $A_f(-3\pi/2) = 1$, $A_f(-2\pi) = 0$

   (b) Since f is 2π-periodic, $A_f(x) = \int_0^x f(t)\,dt = \int_{2\pi}^{x+2\pi} f(t)\,dt$. Now, $A_f(2\pi) = 0$, so $\int_{2\pi}^{x+2\pi} f(x)\,dx = \int_0^{x+2\pi} f(x)\,dx = A_f(x+2\pi)$. Thus, $A_f(x) = A_f(x+2\pi)$ which implies that A_f is 2π-periodic.

   (c) Since f is positive on the interval $[0, \pi]$, A_f is an increasing function on this interval. Thus, $A_f(0) = 0 \le A_f(x) \le A_f(\pi) = 2$ when $0 \le x \le \pi$. Since f is negative on the interval $[\pi, 2\pi]$, A_f is decreasing on this interval. Thus, $A_f(\pi) = 2 \ge A_f(x) \ge A_f(2\pi) = 0$ when $0 \le x \le \pi$. Finally, since A_f is 2π-periodic, we may conclude that $0 \le A_f(x) \le 2$ for all x.

   (d) $A_f(x) = 1 - \cos x$

9. (a) $\displaystyle\int_{\sqrt{\pi/2}}^x f(t)\,dt = \int_0^x f(t)\,dt - \int_0^{\sqrt{\pi/2}} f(t)\,dt = \sin x^2 - \sin(\pi/2) = \sin x^2 - 1$

   (b) $\displaystyle\int_{-\sqrt{3\pi/2}}^x f(t)\,dt = -\int_0^{-\sqrt{3\pi/2}} f(t)\,dt + \int_0^x f(t)\,dt = -\sin(3\pi/2) + \sin x^2 = 1 + \sin x^2$

11. (a) $A_f(5)$ is larger because f is positive on the interval $[1, 5]$

    (b) $A_f(7)$ is larger because f is negative on the interval $[7, 10]$

    (c) $A_f(-2) < A_f(-1) < 0$

    (d) A_f is increasing on the interval $(-2, 6)$

    (e) It is a local maximum because f changes from positive to negative (i.e., A_f changes from increasing to decreasing).

    (f) $A_f(x) = \displaystyle\int_0^x f(t)\,dt = \int_{-2}^x f(t)\,dt - \int_{-2}^0 f(t)\,dt = F(x) + C$ where $C = -\displaystyle\int_{-2}^0 f(t)\,dt$. $C < 0$ because $f(x) > 0$ when $-2 \le x \le 0$.

    (g) $0 < A_f(-1) - A_f(-2) < A_f(0) - A_f(-1) < A_f(1) - A_f(0) < A_f(2) - A_f(1)$

    (h) These values suggest that A_f is concave down on the interval $[3, 8]$—the slopes of the secant lines are decreasing.

13. (a) $G(3)$

    (b) $-G(0)$

    (c) $G(2) - G(-2)$

15. (b) $g(1) = \int_1^1 f(t)\,dt = 0$

    (c) $g(x) = h(x) + \int_1^3 f(t)\,dt$ and $\int_1^3 f(t)\,dt > 0$ since f is positive on the interval $[1, 3]$ ($f(0) = 0$ and $f'(x) \ge 0$ on the interval).

(d) $h(0) < h(4) < g(3) < g(7)$

$h(0) = \int_3^0 f(t)\,dt = -\int_0^3 f(t)\,dt < 0$ since $f(x) > 0$ on the interval $[0, 3)$

$6 < g(3) = \int_1^3 f(t)\,dt < 12$ since $f(1) = 3$, $f(3) = 6$, and $f'(x) \geq 0$ if $1 \leq x \leq 3$

$4 < h(4) = \int_3^4 f(t)\,dt < 6$ since $f(3) = 6$, $f(4) = 4$, and $f'(x) \leq 0$ if $3 \leq x \leq 4$

$g(7) = g(3) + h(4) + \int_4^7 f(t)\,dt$. Now $f(x) \geq 0$ if $4 \leq x \leq 5$ and $f(x) \geq -2$ if $5 \leq x \leq 7$ imply that $-4 \leq \int_4^7 f(t)\,dt$. Thus, $h(4) + \int_4^7 f(t)\,dt > 0$ so $g(3) < g(7)$.

(e) one — $g(1) = 0$. The information given implies that $g'(x) = f(x) \geq 0$ if $0 \leq x \leq 5$ and $g'(x) = f(x) \leq 0$ if $5 \leq x \leq 7$. Thus, g has only one root in the interval $[0.7]$.

(f)

17. (a) $G(x) = H(x) + \int_{-3}^2 f(t)\,dt = H(x) - 2\pi + 1/2$. The values of G and H differ by the signed area of the region bounded by f over the interval $[-3, 2]$.

(b) $(-5, -3)$ and $(1, 5)$

(c) $G'(x) = f(x)$ changes sign (from negative to positive) at $x = 1$.

(d) G achieves its minimum value at $x = -5$; $G(-5) = -3$. G achieves its maximum value at $x = -3$; $G(-3) = 0$.

(e) H achieves its minimum value at $x = 1$; $H(1) = -1/2$. H achieves its maximum value at $x = -3$; $H(-3) = 2\pi - 1/2$.

(f) G is concave down on $(-5, -1)$ and $(3, 5)$

(g) H has points of inflection at $x = -1$ and at $x = 3$.

19. $G(x) - F(x) = \int_b^x g(t)\,dt - \int_a^x f(t)\,dt = \int_b^c g(t)\,dt + \int_c^x g(t)\,dt - \int_a^c f(t)\,dt - \int_c^x f(t)\,dt$

$= G(c) - F(c) + \int_c^x \left(g(t) - f(t)\right) dt \geq 0$ when $x \geq c$.

21. (a) $A_f(x) = \int_{-1/2}^0 f(t)\,dt + \int_0^x f(t)\,dt = \dfrac{\sqrt{3}}{8} + \dfrac{\pi}{12} + \dfrac{1}{2}x\sqrt{1 - x^2} + \dfrac{1}{2}\arcsin x$

(b) Yes.

23. (a) $A_f(x) = x^3/3$

(b) Yes.

5.3   The Fundamental Theorem of Calculus

1. (a) Graph A is the winner. The point is that $F' = f$. Graph C is wrong because it goes down in the middle. Graph B has the wrong direction of concavity.

   (b) The g-graph is just like the F-graph, but *raised* two units vertically (i.e., $g(x) = F(x) + 2$). As always in this section, the point is that F and g have the same derivative, f.

3. (a) $\displaystyle\int_1^4 \left(x + x^{3/2}\right) dx = 199/10$.

   (b) $\displaystyle\int_0^\pi \cos x \, dx = 0$

   (c) $\displaystyle\int_{-2}^5 \frac{dx}{x+3} = \ln 8$

   (d) $\displaystyle\int_0^b x^2 \, dx = b^3/3$

   (e) $\displaystyle\int_1^b x^n \, dx = \frac{b^{n+1}}{n+1} - \frac{1}{n+1}$ $[n \neq -1]$

   (f) $\displaystyle\int_2^{2.001} \frac{x^5}{1000} \, dx = \frac{2.001^6 - 2^6}{6000} \approx 0.00003204$

   (g) $\displaystyle\int_0^{0.001} \frac{\cos x}{1000} \, dx = \frac{\sin 0.001}{1000} \approx 0.000001$

   (h) $\displaystyle\int_0^{\sqrt{\pi}} x \sin\left(x^2\right) dx = -\frac{1}{2}\cos(x^2)\Big]_0^{\sqrt{\pi}} = 1$

5. By the FTC, $\dfrac{d}{dx}\left[\displaystyle\int_a^x f(t)\,dt\right] = f(x)$. The FTC also implies that $\displaystyle\int_a^x \left[\dfrac{d}{dt}f(t)\right] dt = \displaystyle\int_a^x f'(t)\,dt = f(x) - f(a)$.

   Thus if $f(a) \neq 0$, $\dfrac{d}{dx}\left[\displaystyle\int_a^x f(t)\,dt\right] \neq \displaystyle\int_a^x \left[\dfrac{d}{dt}f(t)\right] dt$.

7. (a) $g'(4) = f(4) = 2$

   (b) g is concave up on $[1, 4]$ (i.e., where f is increasing)

   (c) Two — $g(1) = 0$; g also has a root in the interval $[2, 3]$ since $g(2) < 0$ and $g(3) > 0$

   (d) $-1 < g(2) < 0 < g(0) < 1 < g(4)$

   (e) The average value of g' over the interval $[0, 3]$ is $\dfrac{\int_0^3 g'(x)\,dx}{3-0} = \dfrac{g(3) - g(0)}{3}$. Now, $0 < g(0) < 1$ and $0 < g(3) < 1$, so $-1 = 0 - 1 < g(3) - g(0) < 1 - 0 = 1$. Therefore, the average value of g' over the interval $[0, 3]$ is less than 1.

   ALTERNATE SOLUTION: The average value of g' over the interval $[0, 3]$ is $\frac{1}{3}\int_0^3 g'(x)\,dx = \frac{1}{3}\int_0^3 f(x)\,dx$. Since $\int_0^3 f(x)\,dx < 3$, the average value of g' over the interval $[0, 3]$ is less than 1.

9. Let $F(x) = \int_0^x f(t)\,dt = 3x^2 + e^x - \cos x$. Since $F'(x) = f(x) = 6x + e^x + \sin x$, $f(2) = 12 + e^2 + \sin 2$.

11. (a) Let $g(x) = \displaystyle\int_a^x \sqrt[3]{1+t^2}\,dt$. By the FTC, $g'(x) = \sqrt[3]{1+x^2} = f'(x)$. Therefore, since f and g have the same derivative function, $f(x) = g(x) + C$ for some number C. Since $f(1) = 0 = g(1)$, it follows that $C = 0$ (i.e., $f = g$).

    (b) $\displaystyle\int_0^3 \sqrt[3]{1+x^2}\,dx = f(3) - f(0)$

13. The answer is $f(x) = 4x^3 - 3x^2 + 10$. To find it, observe:

    (i) $f'(x) = ax^2 + bx \implies f(x) = ax^3/3 + bx^2/2 + c$

    (ii) $f''(x) = 2ax + b$.

    (iii) $f'(1) = 6 \implies a + b = 6$

    (iv) $f''(1) = 18 \implies 2a + b = 18$

(v) $\int_1^2 f(x)\,dx = 18 = ax^3/3 + bx^2/2 + c \Big]_1^2 = \dfrac{5a}{4} + c + \dfrac{7b}{6}.$

The last three observations provide three equations that must be satisfied by a, b, and c. The solution of this system of three equations in three unknowns is $a = 12$, $b = -6$, and $c = 10$.

15. (a) The demand at time t is $D(t) = 800 - 10t$ where t measures time in months. The production at time t is $P(t) = 900$.

(b) $D(t)$ represents the demand for the product in units/month. Since $D(t)$ is a rate function, $\int_0^t D(s)\,ds$ is the accumulated demand over the interval $[0, t]$.

(c) The inventory at time t is $I(t) = $ starting inventory $+$ total production up to time t $-$ total demand up to time t. Since now $P(t) = 900 - Rt$,

$$
\begin{aligned}
I(t) &= 1680 + \int_0^t (900 - Rs)\,ds - \int_0^t (800 - 10s)\,ds \\
&= (5 - \frac{R}{2})t^2 + 100t + 1680.
\end{aligned}
$$

(d) When $t = 12$ the inventory is $3600 - 72R$; to make this zero, $R = 50$ must be true.

5.4   Approximating Sums: The Integral as a Limit

1. Using 5 equal subintervals, the left sum approximation to $\int_{-5}^{5} g(x)\, dx$ is

$$2\Big(f(-5) + f(-3) + f(-1) + f(1) + f(3)\Big) = 4;$$

the right sum approximation is $2\Big(f(-3) + f(-1) + f(1) + f(3) + f(5)\Big) = 8;$

and, the midpoint sum approximation is $2\Big(f(-4) + f(-2) + f(0) + f(2) + f(4)\Big) = 7.$

Diagrams illustrating each of these sums appear below:

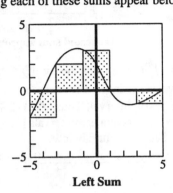

Left Sum

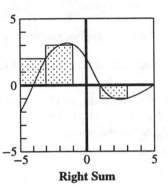

Right Sum

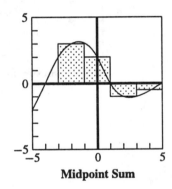

Midpoint Sum

3. $R_{10} = 2 \displaystyle\sum_{k=1}^{10} (2k)^2$

5. left: $\displaystyle\int_{0}^{5} \sqrt[3]{2x}\, dx \approx \frac{5}{10} \sum_{j=0}^{9} \sqrt[3]{2 \cdot j \cdot \frac{5}{10}} = \frac{1}{2} \sum_{j=0}^{9} \sqrt[3]{j}$

   right: $\displaystyle\int_{0}^{5} \sqrt[3]{2x}\, dx \approx \frac{5}{10} \sum_{j=1}^{10} \sqrt[3]{2 \cdot j \cdot \frac{5}{10}} = \frac{1}{2} \sum_{j=1}^{10} \sqrt[3]{j}$

   midpoint: $\displaystyle\int_{0}^{5} \sqrt[3]{2x}\, dx \approx \frac{5}{10} \sum_{j=0}^{9} \sqrt[3]{2 \cdot (j + 0.5) \cdot \frac{5}{10}} = \frac{1}{2} \sum_{j=0}^{9} \sqrt[3]{j + 0.5}$

7. $\dfrac{2}{100} \displaystyle\sum_{k=1}^{100} \sin\left(\frac{2k}{100}\right) \approx \int_{0}^{2} \sin x\, dx = -\cos 2 + \cos 0 \approx 1.41615.$ (This is a right sum approximation to the integral.)

9. (a) The left sum approximation to $\int_{2}^{5} f(x)\, dx$ with $n = 3$ equal subintervals is $f(2) \cdot (3 - 2) + f(3) \cdot (4 - 3) + f(4) \cdot (5 - 4) = 0.21 + 0.28 + 0.36 = 0.85.$

(b) The trapezoid sum approximation is

$$\frac{1}{2}\big(f(2) + f(3)\big) + \frac{1}{2}\big(f(3) + f(4)\big) + \frac{1}{2}\big(f(4) + f(5)\big) = 0.965$$

(c)

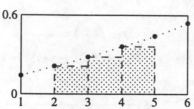
Left sum approximation to $\int_2^5 f(x)\, dx$

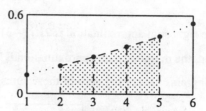
Trapezoid sum approximation to $\int_2^5 f(x)\, dx$

(d) A Riemann sum approximation of an integral $\int_a^b f(x)\, dx$ has the form $\sum_{j=1}^n f(c_j)\, \Delta x_j$ where $a = x_0 < x_1 < \cdots < x_n = b$ is a partition of the interval $[a, b]$ into n subintervals, $\Delta x_j = \big(x_j - x_{j-1}\big)$ is the width of the jth subinterval, and c_j is a point in the jth subinterval. Thus, $f(2) \cdot 2 + f(4) \cdot 3$ is the sum obtained when the interval $[1, 6]$ is partitioned into the two unequal subintervals $[1, 3]$ and $[3, 6]$ and the integrand is evaluated at the points $c_1 = 2$ and $c_2 = 4$. See the picture below.

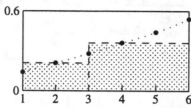
A Riemann sum approximation

11. $\displaystyle \lim_{n \to \infty} \frac{1}{n} \sum_{j=1}^n \left(\frac{j}{n}\right)^3 = \int_0^1 x^3\, dx = \frac{1}{4}$

13. (a) $\displaystyle \sum_{k=1}^4 \frac{5}{k(k+1)}\big(2.3 + (k - 1) \cdot 0.5\big)$

   (b) No. The endpoints of the subintervals are $x_0 = 0$, $x_1 = 5/2$, $x_2 = 10/3$, $x_3 = 15/4$, and $x_4 = 4$. The sampling points are $c_1 = 2.3$, $c_2 = 2.8$, $c_3 = 3.3$, and $c_4 = 3.8$. However, since c_2 and c_3 lie in the same subinterval, the sum is not a Riemann sum.

15. (a) One way is to connect the dots.

   (b) To estimate the total distance traveled:

      (i) A trapezoid approximating sum, 6 subdivisions:

$$T_6 = \frac{1}{6}\,(42/2 + 38 + 36 + 57 + 0 + 55 + 51/2) = 38.75$$

      (ii) A left approximating sum, 6 subdivisions:

$$L_6 = \frac{1}{6}\,(42 + 38 + 36 + 57 + 0 + 55) = 38$$

      (iii) A midpoint approximating sum, 3 subdivisions

$$M_6 = \frac{1}{3}\,(38 + 57 + 55) = 50$$

The trapezoid rule answer might be best.

(c) To plot a plausible distance graph, one might estimate the distances covered over each 10-minute period (using trapezoids) and add them up:

Distance estimates over one hour							
time (min)	0	10	20	30	40	50	60
speed (mph)	42	38	36	57	0	55	51
total distance (miles)	0	6.66	12.83	20.58	25.33	29.92	38.75

Here's a graph of the resulting distance graph:
A possible distance function (min. vs. miles)

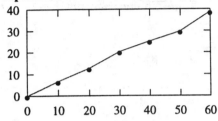

17. (a) $\displaystyle\int_2^{10} f(x)\,dx$

(b) $\displaystyle\int_0^8 f(x)\,dx$

(c) $\displaystyle\int_1^9 f(x)\,dx$

5.5 Approximating Sums: Interpretations and Applications

1. (b) The area is approximately $\dfrac{1}{5}\sum_{i=0}^{4}\left(\dfrac{i}{5}-\dfrac{i^2}{25}\right)=\dfrac{5}{32}=0.15625.$

   (c) Here's the picture:

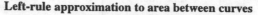

Left-rule approximation to area between curves

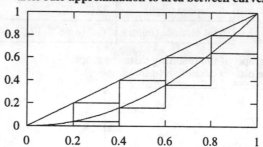

   (d) The area is $\displaystyle\int_0^1 (x-x^2)\,dx = \dfrac{1}{2}x^2 - \dfrac{1}{3}x^3\Big]_0^1 = \dfrac{1}{6}.$

3. (a) $\displaystyle\int_0^1\left(1-\sqrt{1-x}\right)dx + \int_1^2 dx + \int_2^3\left(1-(x-2)\right) = 1/3 + 1 + 1/2 = 11/6$

   (b) $\displaystyle\int_0^1\left(1-\sqrt{1-x}\right)dx + \int_1^2 (2-x)\,dx = 1/3 + 1/2 = 5/6$

   (c) $\displaystyle\int_1^3\left(1-(x-2)\right)dx = \int_1^3 (3-x)\,dx = 3x - \dfrac{1}{2}x^2\Big]_1^3 = 2$

   (d) (a) + (b) = 8/3

5. Area $= \displaystyle\int_{-1}^1 (1-x^4)\,dx = \dfrac{8}{5}.$

7. Area $= \displaystyle\int_0^1 (x^2 - x^3)\,dx = \dfrac{1}{12}$

9. Area $= \displaystyle\int_{-3}^2 [(2-y)-(y^2-4)]\,dy = \int_{-4}^0 2\sqrt{x+4}\,dx + \int_0^5\left[(2-x)+\sqrt{x+4}\right]dx = \dfrac{125}{6}$

11. Area $= \displaystyle\int_0^1 (\sqrt{x}-x^2)\,dx = \dfrac{1}{3}$

13. Area $= \displaystyle\int_{-1}^1\left[(2-x^2)-\dfrac{9}{4x^2+5}\right]dx = \dfrac{10}{3} - \dfrac{9}{\sqrt{5}}\arctan\left(\dfrac{2}{\sqrt{5}}\right) \approx 0.39624$

15. Area $= \displaystyle\int_0^1 e^x\,dx = e - 1$

6.1   Antiderivatives: The Idea

1. $\int \left(3x^5 + 4x^{-2}\right) dx = \frac{1}{2}x^6 - 4/x + C$

3. $\int \frac{dx}{4\sqrt{1-x^2}} = \frac{1}{4}\arcsin x + C$

5. $\int 3e^{4x}\, dx = \frac{3}{4}e^{4x} + C$

7. $\int 4\sec^2(3x)\, dx = \frac{4}{3}\tan(3x) + C$

9. $\int (x+1)^2 \sqrt[3]{x}\, dx = \int \left(x^{7/3} + 2x^{4/3} + x^{1/3}\right) dx = \frac{3}{10}x^{10/3} + \frac{6}{7}x^{7/3} + \frac{3}{4}x^{4/3} + C$

11. $\int \frac{(1-x)^3}{\sqrt{x}} = \int \left(x^{-1/2} - 3x^{1/2} - x^{5/2}\right) dx = 2x^{1/2} - 2x^{3/2} + \frac{6}{5}x^{5/2} - \frac{2}{7}x^{7/2} + C$

13. **False.** $\left(e^{-\cos x} + C\right)' = e^{-\cos x}\sin x \neq e^{\sin x}$

15. **True.** $\left(\tan x + C\right)' = \sec^2 x = 1 + \tan^2 x$

17. $\left(x\arctan x - \frac{1}{2}\ln\left(1+x^2\right) + C\right)' = \arctan x + \frac{x}{1+x^2} - \frac{1}{2}\cdot\frac{2x}{1+x^2} = \arctan x$

19. $\left(\ln|\sec x| + C\right)' = \frac{\sec x \tan x}{\sec x} = \tan x$

21. $\left(\frac{1}{2}\ln\left|\frac{1+x}{1-x}\right| + C\right)' = \frac{1}{2}\cdot\frac{1-x}{1+x}\cdot\frac{2}{(1-x)^2} = \frac{1}{1-x^2}$

23. (a) $\begin{aligned}\int \sin^2 x\, dx &= \int \frac{1}{2}(1 - \cos(2x))\, dx \\ &= \frac{1}{2}\left(\int 1\, dx - \int \cos(2x)\, dx\right) \\ &= \frac{1}{2}\left(x - \frac{1}{2}\sin(2x)\right) + C \\ &= \frac{x}{2} - \frac{1}{4}\sin(2x) + C\end{aligned}$

   (b) $\sqrt{1 - \cos(2x)} \neq \sqrt{2}\sin x$ when $\pi < x < 2\pi$.

25. $\frac{1}{2}\sin(2x) = \frac{1}{2}\cdot 2\cos x \sin x = \cos x \sin x$

27. $\int \frac{x^2}{1+x^2}\, dx = x - \arctan x + C$

29. $\int \frac{x-1}{x+1}\, dx = x - 2\ln|x+1| + C$

31. $\int \tan^2 x\, dx = \tan x - x + C$

33. $\int \frac{dx}{1+9x^2} = \frac{1}{3}\arctan(3x) + C$

35. $\int \frac{2}{\sqrt{1-9x^2}}\, dx = \frac{2}{3}\arcsin(3x) + C$

6.2   Antidifferentiation by Substitution

1. $\displaystyle\int (4x+3)^{-3}\,dx = -\frac{1}{8(4x+3)^2} + C$

3. $\displaystyle\int e^{\sin x}\cos x\,dx = e^{\sin x} + C$

5. $\displaystyle\int \frac{\arctan x}{1+x^2}\,dx = \frac{1}{2}(\arctan x)^2 + C$

7. $\displaystyle\int \frac{e^{1/x}}{x^2}\,dx = \int \frac{e^{x^{-1}}}{x^2}\,dx = -e^{x^{-1}} + C = -e^{1/x} + C$

9. $\displaystyle\int \frac{e^x}{1+e^{2x}}\,dx = \arctan(e^x) + C$

11. $u = x^2,\ a = 4,\ b = 1;\quad \displaystyle\int_{-2}^{1} \frac{x}{1+x^4}\,dx = \frac{1}{2}\int_{4}^{1} \frac{1}{1+u^2}\,du = (\pi/4 - \arctan 4)/2 \approx -0.27021$

13. $u = 2x^2 + 1,\ a = 1,\ b = 19;\quad \displaystyle\int_{0}^{3} \frac{x}{(2x^2+1)^3}\,dx = \frac{1}{4}\int_{1}^{19} u^{-3}\,du = \frac{45}{361}$

15. $\displaystyle\left(\frac{1}{2}\arctan(x^2) + C\right)' = \frac{1}{2}\frac{2x}{1+\left(ix^2\right)^2} = \frac{x}{1+x^4}$

17. (a) If $u = 1 + \sqrt{x}$, then $du = \frac{1}{2}x^{-1/2}\,dx$ and $2(u-1)\,du = dx$. Thus,
$\displaystyle\int \frac{dx}{1+\sqrt{x}} \to \int \frac{2(u-1)\,du}{u} = 2u - 2\ln|u| + C \to 2(1+\sqrt{x}) - 2\ln(1+\sqrt{x}) + C$

    (b)

19. If $u = a + b/x$, then $du = -\frac{b}{x^2}\,dx$. Therefore,
$\displaystyle\int \frac{dx}{ax^2+bx} = \int \frac{dx}{x^2(a+b/x)} \to -\frac{1}{b}\int \frac{du}{u} = -\frac{1}{b}\ln|u| + C \to -\frac{1}{b}\ln|a+b/x| + C =$
$-\frac{1}{b}\ln|ax+b| + \frac{1}{b}\ln|x| + C.$

21. If $\int_0^{12} g(x)\,dx = \pi$, then $\int_0^4 g(3x)\,dx = \frac{\pi}{3}$. **Reason:** Make the substitutions $u = 3x$ and $du = 3\,dx$ in the second integral: $\displaystyle\int_0^4 g(3x)\,dx = \frac{1}{3}\int_0^4 3\cdot g(3x)\,dx = \frac{1}{3}\int_0^{12} g(u)\,du = \frac{1}{3}\pi$

23. Let $u = 2x + 3$. Then, $\displaystyle\int \cos(2x+3)\,dx = \frac{1}{2}\int \cos u\,du = \frac{1}{2}\sin u + C = \frac{1}{2}\sin(2x+3) + C.$

25. Let $u = 3x - 2$. Then $du = 3\,dx$ and $\frac{1}{3}\,du = dx$. Therefore, $\displaystyle\int (3x-2)^4\,dx \to \frac{1}{3}\int u^4\,du = \frac{1}{15}u^5 + C \to$
$\frac{1}{15}(3x-2)^5 + C.$

27. Let $u = 1 + x^4$. Then, $\displaystyle\int \frac{2x^3}{1+x^4}\,dx = \frac{\ln(1+x^4)}{2} + C.$

29. Let $u = 1 - 2x$. Then, $du = -2\,dx$ and $-\frac{1}{2}\,du = dx$. Therefore, $\displaystyle\int \frac{dx}{1-2x} \to -\frac{1}{2}\int \frac{du}{u} = -\frac{1}{2}\ln|u| + C \to$
$-\frac{1}{2}\ln|1-2x| + C.$

31. Let $u = 3 - 2x$. Then $x = (3-u)/2$, $du = -2\,dx$, and $-\frac{1}{2}\,du = dx$. Therefore,

$$\int x\sqrt{3-2x}\,dx \;\to\; -\frac{1}{4}\int (3-u)u^{1/2}\,du = -\frac{1}{4}\int \left(3u^{1/2} - u^{3/2}\right)du$$

$$= \tfrac{1}{10}u^{5/2} - \tfrac{1}{2}u^{3/2} + C$$

$$\to \tfrac{1}{10}(3-2x)^{5/2} - \tfrac{1}{2}(3-2x)^{3/2} + C.$$

33. Let $u = 1 + x^{-1}$. Then, $\displaystyle\int \frac{\sqrt{1+x^{-1}}}{x^2}\,dx = -\frac{2}{3}\left(1+x^{-1}\right)^{3/2} + C.$

35. Let $u = x^4 - 1$. Then, $\displaystyle\int x^3\left(x^4 - 1\right)^2 dx = \frac{1}{12}\left(x^4 - 1\right)^3 + C.$

37. Let $u = x^3$. Then, $\displaystyle\int \frac{x^2}{1+x^6}\,dx = \frac{1}{3}\arctan\left(x^3\right) + C.$

39. Let $u = 4x^3 + 5$. Then, $\displaystyle\int x^2\sqrt{4x^3+5}\,dx = \frac{1}{18}\left(4x^3+5\right)^{3/2} + C.$

41. $\displaystyle\int \frac{x+4}{x^2+1}\,dx = \int \frac{x}{x^2+1}\,dx + 4\int \frac{dx}{x^2+1} = \int \frac{x}{x^2+1}\,dx + 4\arctan x = \frac{1}{2}\ln\left(1+x^2\right) + 4\arctan x + C.$
(The substitution $u = 1 + x^2$ is used in the last step.)

43. Let $u = x^2 + 3x + 5$. Then, $\displaystyle\int \frac{2x+3}{\left(x^2+3x+5\right)^4}\,dx = -\frac{1}{3}\left(x^2+3x+5\right)^3 + C.$

45. Let $u = 3x^2 + 6x + 5$. Then, $\displaystyle\int \frac{x+1}{\sqrt[3]{3x^2+6x+5}}\,dx = \frac{1}{4}\left(3x^2+6x+5\right)^{2/3} + C.$

47. Let $u = 2e^x + 3$. Then, $\displaystyle\int \frac{e^x}{(2e^x+3)^2}\,dx = -\frac{1}{2}\left(2e^x+3\right)^{-1} + C.$

49. Let $u = x + 1$. Then, $\displaystyle\int \frac{2x+3}{(x+1)^2}\,dx = \int \frac{2(u-1)+3}{u^2}\,du = 2\ln|x+1| - (x+1)^{-1} + C.$

51. Let $u = \cos x$. Then, $\displaystyle\int \tan x\,dx = -\ln|\cos x| + C.$

53. Let $u = \arcsin x$. Then, $\displaystyle\int \frac{\arcsin x}{\sqrt{1-x^2}}\,dx = \frac{1}{2}(\arcsin x)^2 + C.$

55. Let $u = 3x^2 + 4$. Then, $\displaystyle\int \frac{5x}{3x^2+4}\,dx = \frac{5}{6}\ln\left(3x^2+4\right) + C.$

57. Let $u = \tan x$. Then, $\displaystyle\int \frac{\sec^2 x}{\sqrt{1-\tan^2 x}}\,dx = \arcsin(\tan x) + C.$

59. Let $u = x^2$. Then, $\displaystyle\int x\tan\left(x^2\right)dx = \frac{1}{2}\int \tan u\,du = \frac{1}{2}\int \frac{\sin u}{\cos u}\,du = -\frac{1}{2}\ln|\cos u| + C =$
$-\frac{1}{2}\ln\left|\cos\left(x^2\right)\right| + C.$

61. Let $u = \sin x$. Then, $\displaystyle\int \frac{\cos x}{\sin^4 x}\,dx = -\frac{1}{3\sin^3 x} + C = -\tfrac{1}{3}\csc^3 x + C.$

63. Let $u = \ln(\cos x)$. Then $du = -\frac{\sin x}{\cos x} dx = -\tan x\, dx$. Therefore, $\int \ln(\cos x) \tan x\, dx \to -\int u\, du$
$= -\frac{1}{2}u^2 + C \to -\frac{1}{2}\left(\ln(\cos x)\right)^2 + C.$

65. Let $u = sqrtx + 2$. Then $du = \frac{1}{2}x^{-1/2}\, dx$ and $dx = 2(u-2)du$. Therefore, $\displaystyle\int \frac{dx}{\sqrt{x}\left(\sqrt{x}+2\right)^3} \to$
$2\displaystyle\int \frac{du}{u^3} = -\frac{1}{u^2} + C \to -\left(\sqrt{x}+2\right)^{-2} + C.$

67. Let $u = 1 + \sqrt{x}$. Then $du = \frac{1}{2}x^{-1/2}\, dx$ and $2(u-1)\, du = dx$. Therefore, $\displaystyle\int \sqrt{1+\sqrt{x}}\, dx \to$
$\displaystyle\int u^{1/2}\cdot 2(u-1)\, du = \frac{4}{5}u^{5/2} - \frac{2}{3}u^{3/2} + C \to \frac{4}{5}(1+\sqrt{x})^{5/2} - \frac{2}{3}(1+\sqrt{x})^{3/2} + C.$

69. First, note that $\displaystyle\int \frac{e^{\tan x}}{1 - \sin^2 x} dx = \int \frac{e^{\tan x}}{\cos^2 x} dx = \int \sec^2 x\, e^{\tan x}\, dx$. Now, let $u = \tan x$. Then,
$\displaystyle\int \frac{e^{\tan x}}{1 - \sin^2 x} dx = e^{\tan x} + C.$

71. Let $u = 8 - x$. Then, $\displaystyle\int_{-19}^{8} \sqrt[3]{8-x}\, dx = -\frac{3}{4}(8-x)^{4/3}\Big]_{-19}^{8} = \frac{243}{4}$

73. Let $u = \ln x$. Then, $\displaystyle\int_{e}^{4e} \frac{dx}{x\sqrt{\ln x}} = 2\sqrt{\ln x}\,\Big]_{e}^{4e} = 2\sqrt{1 + 2\ln 2} - 2$

75. Let $u = \cos x$. Then, $\displaystyle\int_{-\pi/2}^{\pi} e^{\cos x} \sin x\, dx = -e^{\cos x}\Big]_{-\pi/2}^{\pi} = 1 - e^{-1}$

77. Let $u = \pi - x$. Then $x = \pi - u$ and $du = -dx$. Therefore,

$$\int_0^\pi x f(\sin x)\, dx = -\int_\pi^0 (\pi - u) f\big(\sin(\pi - u)\big)\, du$$
$$= \int_0^\pi (\pi - u) f\big(\sin(\pi - u)\big)\, du$$
$$= \int_0^\pi (\pi - u) f(\sin u)\, du$$
$$= \pi \int_0^\pi f(\sin u)\, du - \int_0^\pi u f(\sin u)\, du$$

This implies that $2\displaystyle\int_0^\pi x f(\sin x)\, dx = \pi \int_0^\pi f(\sin x)\, dx$ or, equivalently, that
$\displaystyle\int_0^\pi x f(\sin x)\, dx = \frac{\pi}{2}\int_0^\pi f(\sin x)\, dx.$

79. Let $u = 1 - x$. Then, $\displaystyle\int_0^1 x^n (1-x)^m\, dx = \int_1^0 (1-u)^n u^m\, (-du) = \int_0^1 u^m (1-u)^n\, du.$

81. First, note that $\displaystyle\int \frac{dx}{\sqrt{x}+\sqrt[3]{x}} = \int \frac{dx}{\sqrt[3]{x}\left(\sqrt[6]{x}+1\right)}$. Now, let $u^6 = x$ and $w = u + 1$ to obtain

$$
\begin{aligned}
\int \frac{dx}{\sqrt{x}+\sqrt[3]{x}} &= 6\int \frac{u^5}{u^2(u+1)} = 6\int \frac{u^3}{u+1} \\
&= 6\int \frac{(w-1)^3}{w}\,dw = 6\int \left(w^2 - 3w + 3 - w^{-1}\right) dw \\
&= 2w^3 - 9w^2 + 18w - 6\ln w + C \\
&= 2x^{1/2} - 3x^{1/3} + 6x^{1/6} - 6\ln\left(x^{1/6}+1\right) + C.
\end{aligned}
$$

83. $\sqrt{1 - \sin^2 x} = -\cos x$ if $\pi/2 \le x \le \pi$. The "proof" incorrectly replaces $\sqrt{1 - \sin^2 x}$ by $\cos x$ for all x in the interval $[0, \pi]$.

6.3 Integral Aids: Tables and Computers

1. $\displaystyle\int \frac{dx}{3+2e^{5x}} = \frac{1}{3}x - \frac{1}{15}\ln\left(3+2e^{5x}\right).$ [Use formula #58.]

3. $\displaystyle\int \frac{dx}{x^2(3-x)} = -\frac{1}{3x} - \frac{1}{9}\ln\left|\frac{3-x}{x}\right| = \frac{1}{9}\ln\left|\frac{x}{3-x}\right| - \frac{1}{3x}.$ [Use formula #24.]

5. $\displaystyle\int \tan^3(5x)\,dx = \frac{1}{10}\tan^2(5x) - \frac{1}{5}\ln|\cos(5x)|.$ [Use formulas #50 and #7.]

7. $\displaystyle\int x\sin(2x)\,dx = \frac{1}{4}\sin(2x) - \frac{x}{2}\cos(2x).$ [Use formula #46.]

9. $\displaystyle\int e^{2x}\cos(3x)\,dx = \frac{e^{2x}}{13}\left(2\cos(3x)+3\sin(3x)\right).$ [Use formula #55.]

11. $\displaystyle\int \frac{dx}{4-x^2}\,dx = \frac{1}{4}\ln\left|\frac{2+x}{2-x}\right|.$ [Use formula #30.]

13. $\displaystyle\int \frac{4x+5}{(2x+3)^2}\,dx = 4\int \frac{x}{(2x+3)^2}\,dx + 5\int \frac{dx}{(2x+3)^2} = \frac{1}{2(2x+3)} + \ln|2x+3|.$ [Use formulas #22 and #19.]

15. $\displaystyle\int \frac{dx}{4x^2-1} = \frac{1}{4}\ln\left|\frac{2x-1}{2x+1}\right|.$ [Use formula #32.]

17. $\displaystyle\int \frac{x+2}{2+x^2}\,dx = \int \frac{x}{2+x^2}\,dx + 2\int \frac{dx}{2+x^2} = \frac{1}{2}\ln\left(x^2+2\right) + \sqrt{2}\arctan(x\sqrt{2}/2).$
[Use formulas #35 and #31.]

19. $\displaystyle\int \frac{5}{4x^2+20x+16}\,dx = \frac{5}{4}\int \frac{dx}{x^2+5x+4}\,dx = \frac{5}{4}\int \frac{dx}{(x+5/2)^2-9/4}\,dx = \frac{5}{12}\ln\left|\frac{x+1}{x+4}\right|.$ [Use formula #32.]

21. Making the substitution $u = x^2$ together with formula #47 (or #49),

$$\int x^3\cos(x^2)\,dx \to \frac{1}{2}\int u\cos u\,du = \frac{1}{2}(u\sin u + \cos u) \to \frac{1}{2}\left(x^2\sin\left(x^2\right) + \cos\left(x^2\right)\right)$$

23. $\displaystyle\int \frac{dx}{(x^2+3x+2)^2} = \int \frac{dx}{((x+3/2)^2-1/4)^2} = -\frac{2x+3}{x^2+3x+2} - 2\ln\left|\frac{x+1}{x+2}\right|.$ [Use formula #33.]

25. $\displaystyle\int \frac{e^x}{e^{2x}-2e^x+5}\,dx \to \int \frac{du}{u^2-2u+5} = \int \frac{du}{(u-1)^2+4} \to \int \frac{dw}{w^2+4} = \frac{1}{2}\arctan(w/2) \to$
$\frac{1}{2}\arctan\left(\frac{e^x-1}{2}\right) + C.$
[Use formulas #13 or #31.]

27. $\displaystyle\int \sqrt{x^2+4x+1}\,dx = \int \sqrt{(x+2)^2-3}\,dx = \frac{1}{2}\left((x+2)\sqrt{x^2+4x+1} - 3\ln\left|x+2+\sqrt{x^2+4x+1}\right|\right).$
[Complete the square, then use formula #36.]

29. $\displaystyle\int \frac{\cos x \sin x}{(\cos x - 4)(3\cos x + 1)}\, dx \;=\; \int \frac{\cos x \sin x}{3\cos^2 x - 11\cos x - 4}\, dx$

$\displaystyle\rightarrow \; -\int \frac{u}{3u^2 - 11u - 4}\, du$

$\displaystyle= \; -\frac{1}{3}\int \frac{u}{u^2 - 11u/3 - 4/3}\, du = -\frac{1}{3}\int \frac{u}{(u - 11/6)^2 - 169/36}\, du$

$\displaystyle\rightarrow \; -\frac{1}{3}\int \frac{w + 11/6}{w^2 - (13/6)^2}\, dw$

$\displaystyle= \; -\frac{4}{13}\ln|w - 13/6| - \frac{1}{39}\ln|w + 13/6|$

$\displaystyle\rightarrow \; -\frac{4}{13}\ln|u - 4| - \frac{1}{39}\ln|u + 1/3|$

$\displaystyle\rightarrow \; -\frac{4}{13}\ln|\cos x - 4| - \frac{1}{39}\ln|\cos x + 1/3|$

[Use formula #32.]

31. $\displaystyle\int x \sin(3x + 4)\, dx \;\rightarrow\; \frac{1}{9}\int (u - 4)\sin u\, du$

$\displaystyle= \; \frac{1}{9}\left(\sin u - u\cos u + 4\cos u\right)$

$\displaystyle\rightarrow \; \frac{1}{9}\left(\sin(3x + 4) - (3x + 4)\cos(3x + 4) + 4\cos(3x + 4)\right)$

[Use formula #46 or #48.]

7.1   The Idea of Approximation

1. (a) $I = \int_1^4 \dfrac{dx}{\sqrt{x}} = 2\sqrt{x}\,\Big]_1^4 = 4 - 2 = 2$

    (b) $L_3 = \dfrac{4-1}{3}\left(f(1) + f(2) + f(3)\right) = \dfrac{1}{\sqrt{1}} + \dfrac{1}{\sqrt{2}} + \dfrac{1}{\sqrt{3}} \approx 2.28446;$

    $R_3 = \dfrac{4-1}{3}\left(f(2) + f(3) + f(4)\right) = \dfrac{1}{\sqrt{2}} + \dfrac{1}{\sqrt{3}} + \dfrac{1}{\sqrt{4}} \approx 1.78446$

    (c) $|I - L_3| \approx 0.28446;\ |I - R_3| \approx 0.21554$

    (d) Yes — the actual approximation errors are less than $|f(4) - f(1)|\,\dfrac{4-1}{3} = \left|\dfrac{1}{\sqrt{4}} - 1\right| = \dfrac{1}{2}$ (the bound given in Theorem 1).

    (e) The approximation error made by T_3 is less than that allowed by Theorem 1: $T_3 = (L_3 + R_3)/2 \approx 2.03446$ so $|I - T_3| \approx 0.03446 \le 0.25$ (the error bound for the trapezoid rule estimate).

    (f) The value of n must be chosen so that $|f(4) - f(1)|\,\dfrac{4-1}{n} = \dfrac{3}{2n} \le 0.005$. This will be true if $n \ge 300$.

3. (a) A reasonable guess is around 1.

    (b) $L_{20} \approx 0.961197;\ R_{20} \approx 0.811215$. They overestimate and underestimate, respectively, since the function is decreasing on the interval of integration.

    (c) For L_n and R_n, the theorem says we're OK if $\dfrac{|f(3) - f(0)| \cdot (3 - 0)}{n} \approx \dfrac{3}{n} \le 0.005$, i.e., if $n \ge 600$.

5. For $\int_1^2 x^2\,dx$ we need $n \ge 600{,}000$.

7. For $\int_1^2 x^{-1}\,dx$ we need $n \ge 100{,}000$.

9. $n \ge 800{,}000$

11. $n \ge 600{,}000$

13. $n \ge 153{,}636$

15. $n \ge 300{,}000$

17. $n \ge 50{,}000$

19. (a) No. Since f is decreasing over the interval of integration, L_n overestimates I for any n.

    (b) By the theorem, $|I - L_{16}| \le \dfrac{|f(1) - f(0)| \cdot (1 - 0)}{16} = \dfrac{3}{16}.$

    (c) The idea is that since f is monotone, we can find the *difference* between L_{16} and R_{16}. That is,

    $$|L_{16} - R_{16}| = \dfrac{|f(1) - f(0)| \cdot (1 - 0)}{16} = \dfrac{3}{16}.$$

    Moreover, since f is decreasing, we know that $L_n > R_n$, so $R_{16} = L_{16} + \dfrac{3}{16} = 5.5047.$

    (d) By definition, $T_{16} = R_{16} + L_{16})/2 = (5.3172 + 5.5047)/2 = 5.41095.$ By the theorem, $|I - T_{16}| \le \dfrac{|f(1) - f(0)| \cdot (1 - 0)}{2 \cdot 16} = \dfrac{3}{32}.$

21. A picture similar to that in the proof of Theorem 1 makes this clear.

23. The average value of f over the interval $[1, 5]$ is $\dfrac{1}{4}\displaystyle\int_1^5 f(x)\,dx$. Therefore, $T_n/4$ will estimate the average value of f over the interval $[1, 5]$ within 0.01 if n is chosen such that $\left|\int_1^5 f(x)\,dx - T_n\right| \le 0.04$. Theorem 1 guarantees that this inequality holds for all $n \ge 22$. Thus, $T_{22}/4 \approx 0.9028$ approximates the average value of f over the interval $[1, 5]$ within 0.01.

25. Notice that the integrand is *not* monotone over the interval $[0, 2]$. Thus before using Theorem 1 we should break the interval of integration two pieces, on each of which the integrand *is* monotone. A look at the graph shows that the function is *increasing* on $[0, \sqrt{\pi/2}]$ and *decreasing* on $[\sqrt{\pi/2}, 2]$. Thus we can approximate answers on each of these intervals, say with error less than 0.005, and add the results. The respective answers are about

$$\int_0^{\sqrt{\pi/2}} \sin\left(x^2\right)\,dx \approx 0.549276; \qquad \int_{\sqrt{\pi/2}}^2 \sin\left(x^2\right)\,dx \approx 0.255652;$$

a good estimate to I, therefore, is around 0.80 or 0.81. (More advanced methods can be used to show that the "true" answer is about 0.80485.)

27. The statement **must** be true. The condition $f'(x) > 0$ for all x in $[3, 8]$ means that f is increasing, so *all* left sums underestimate I.

29. The statement **may** be true. $|I - L_{1000}| \le |-4 - 2|\,\dfrac{5}{1000} = 0.03$.

31. The statement **must** be true. Because f is monotone, Theorem 1 applies. It says that if f is monotone on $[3, 8]$, $f(3) = 5$, and $f(8) = 1$, then $|I - T_{1000}| \le |f(8) - f(3)|\dfrac{5}{2000} = \dfrac{20}{2000} = 0.01 < 0.05$.

33. The statement **may** be true. If the function f is not monotone, it may happen that the actual approximation error made by R_{20} is greater than that of R_{10}.

35. Recall that $T_n = (L_n + R_n)/2$. Into this, substitute the expression for R_n derived in the previous exercise.

37. (a) $T_{10} = \frac{1}{2}(L_{10} + R_{10}) = 9.495$. Using equation 6.1.1, $|I - T_{10}| \le \frac{1}{2}|R_{10} - L_{10}| = 0.0818$.

    (b) Since f is an increasing function on the interval of integration, $L_{10} \le I \le R_{50}$ is true. Thus, the midpoint of the interval from L_{10} to R_{50}, $\frac{1}{2}(L_{10} + R_{50})$, cannot be further from I than one-half the length of the interval, $R_{50} - L_{10}$.

39. As the picture in the proof of Theorem 1 shows, the "exact" integral I must lie somewhere *between* L_n and R_n, i.e., in an interval of length

$$|R_n - L_n| = |f(b) - f(a)|\,\Delta x = |f(b) - f(a)|\,\frac{(b-a)}{n}.$$

Since T_n is the *midpoint* of this same interval, I must lie within a distance of *half* the interval's width from T_n. (Draw L_n, R_n, and T_n on a number line to understand all this.)

41. (a) By the Theorem's formula, $|I - L_4| \le \dfrac{9 \cdot 11}{n}$; thus we want $\dfrac{9 \cdot 11}{n} \le 0.005$, or $n \ge 19{,}800$. A lot!

    (b) By the Theorem's formula, $|I - L_4| \le \dfrac{8 \cdot 1}{n}$; thus we want $\dfrac{8}{n} \le 0.004$, or $n \ge 2{,}000$.

    (c) By the Theorem's formula, $|I - L_4| \le \dfrac{1 \cdot 10}{n}$; thus we want $\dfrac{10}{n} \le 0.001$, or $n \ge 10{,}000$.

    (d) Adding the estimates produced in the previous two parts does the trick.

    (e) The estimate in part (d) requires significantly less computational effort.

43. The point of (iv), in each case, is to see that the ACTUAL errors committed are no more than the theoretical bounds guaranteed by Theorem 1. Each exercise has many parts; here they are, tabulated.

(a) $I = \int_1^2 x^2\, dx = 7/3$.

n	4	8	16	32	64	128
L_n	1.9688	2.1485	2.2403	2.2867	2.3103	2.3220
$I - L_n$	0.3645	0.1848	0.0930	0.0466	0.0230	0.0113
Thm 1 bound	0.7500	0.3750	0.1875	0.0937	0.0468	0.0234

(b) $I = \int_1^4 \sqrt{x}\, dx = 14/3$.

n	4	8	16	32	64	128
L_n	4.2802	4.4764	4.5724	4.6199	4.6435	4.6560
$I - L_n$	0.3865	0.1903	0.0943	0.0468	0.0232	0.0107
Thm 1 bound	0.75000	0.37500	0.18750	0.0937	0.0468	0.0234

(c) $I = \int_1^2 x^{-1}\, dx = \ln 2 \approx 0.6932$.

n	4	8	16	32	64	128
L_n	0.75953	0.72538	0.70900	0.70103	0.69709	0.69514
$I - L_n$	−0.0664	−0.0322	−0.01585	−0.0079	−0.0039	−0.0020
Thm 1 bound	−0.1250	−0.0625	−0.0312	−0.0156	−0.0078	−0.0039

(d) $I = \int_2^3 \sin x\, dx = -\cos 3 + \cos 2 \approx 0.57384$.

n	4	8	16	32	64	128
L_n	0.6669	0.6211	0.5976	0.5858	0.5798	0.5768
$I - L_n$	−0.0930	−0.0472	−0.0238	−0.0120	−0.0600	−0.0301
Thm 1 bound	−0.19205	−0.09602	−0.04801	−0.02400	−0.01200	−0.00600

45. The answers can be read from the tables found in Exercise 43. The point is to see that for the Left rule, *doubling* n roughly *halves* the error committed. Thus the error committed with $n = 256$ should be roughly $1/2$ that committed with $n = 128$.

7.2   More on Error: Left and Right Sums and the First Derivative

1. $I = 1$; $L_{10} = 1.000$; $R_{10} = 1.000$; $|I - L_{10}| = 0$; $|I - R_{10}| = 0$; $K_1 = 0$; $\dfrac{K_1(b-a)^2}{2n} = 0$

3. $I = 2.333$; $L_{10} = 2.185$; $R_{10} = 2.485$; $|I - L_{10}| = 0.148$; $|I - R_{10}| = 0.152$; $K_1 = 4$; $\dfrac{K_1(b-a)^2}{2n} = 0.2000$

5. $I = 0.6931$; $L_{10} = 0.7187$; $R_{10} = 0.6687$; $|I - L_{10}| = 0.0256$; $|I - R_{10}| = 0.0244$;

   $K_1 = 1$; $\dfrac{K_1(b-a)^2}{2n} = 0.0500$

7. The condition $-4 \le f'(x) \le 3$ if $1 \le x \le 2$ implies that K_1 must be a number greater than $|-4| = 4$. Thus, Theorem 2 guarantees that $|I - L_n| \le 4/(2n)$.

9. For $\int_0^2 \sin\left(x^2\right) dx$, $K_1 = 4$ works, so any $n \ge 4 \cdot 2^2 \cdot 100 = 1600$ will do.

11. For $\int_1^{10} \sin(1/x)\, dx$, $K_1 = \cos 1$ works so any $n \ge \cos 1 \cdot 9^2 \cdot 100 \approx 4376.45$ will do.

13. (a) $I = \pi$ because the integral gives the area of the northeast quadrant of a circle of radius 2.

    (b) $L_{10} \approx 3.3045$; $|I - L_{10}| \approx 0.1629$.

    (c) Theorem 2 doesn't give a good bound here because we can't compute K_1—$f'(x)$ is unbounded on the interval $(0, 2)$.

    (d) Theorem 1 says that

    $$|L_{10} - I| \le \frac{|f(2) - f(0)| \cdot 2}{10} = 0.4.$$

    This number is larger, as it should be, than the actual error committed.

15. The information given implies that f is increasing and concave down on the interval $[a, b]$. A sketch shows that the approximation error made by L_n includes all of the area corresponding to the approximation error made by T_n and more.

    An algebraic proof of this result is also possible. Since f is (strictly) increasing and (strictly) concave down on the interval of integration, $L_n < I < R_n$ and $I - T_n < 0$. Thus,

    $$(I - L_n) - (I - T_n) = T_n - L_n = \tfrac{1}{2}(R_n - L_n) > 0$$

    which implies that $|I - T_n| < |I - L_n|$.

17. (a) From the graph it is apparent that $\left|f'(x)\right| \le 9$ if $0 \le x \le 4$. Thus, the error bound inequality

    $$|I - L_n| \le \frac{9 \cdot 4^2}{2n} \le 0.0001$$

    implies that any value of $n \ge 720{,}000$ will do the trick.

    (b) **No.** Since the value of $f'(x)$ is negative for all x in the interval of integration, f is decreasing over this interval. It follows that R_n *under*estimates I (i.e., $R_n < I$).

    (c) Since $F' = f$, we need a bound on $|f(x)|$ if $2 \le x \le 4$. From the graph, $|f'(x)| \le 4$ on $2 \le x \le 4$ so $f(2) = 0 \implies |f(x)| \le 4(x - 2)$ if $2 \le x \le 4$. Therefore, $|f(x)| \le 8$ over the interval $[2, 4]$ and so $|L_n - \int_2^4 F(x)\,dx| \le \frac{8 \cdot 2^2}{2n} \le 0.01$ if $n \ge 1600$.

19. Use a picture similar to that in Example 3.

21. (a) Since f is increasing, the approximation error made on each subinterval is nonnegative. Furthermore, this approximation error is less than or equal to $K_1 (\Delta x)^2/2$ (the area of a triangle of width Δx and height $K_1 \Delta x$). Since there are n subintervals,

$$0 \leq I - L_n \leq \sum_{i=1}^{n} \frac{K_1(\Delta x)^2}{2} = \sum_{i=1}^{n} \frac{K_1(b-a)^2}{2n^2} = \frac{K_1(b-a)^2}{2n}.$$

(b) Since f is increasing, the approximation error made on each subinterval is nonpositive. Now, an argument similar to that in part (a) leads to the desired result.

(c) When $f(x) = x$,

$$L_n = \left(\frac{b-a}{n}\right) \sum_{i=0}^{n-1} (a+i(b-a)/n) = \left(\frac{b-a}{n}\right) (na+(n-1)(b-a)) = a(b-a)+\frac{(b-a)^2}{2}-\frac{(b-a)^2}{2n}.$$

Therefore, $I - L_n = \left(\frac{b^2}{2} - \frac{a^2}{2}\right) - a(b-a) - \frac{(b-a)^2}{2} + \frac{(b-a)^2}{2n} = \frac{(b-a)^2}{2n}.$

(d) Taking the *average* of L_n and R_n will produce a better estimate: $(L_n + R_n)/2$. This is the Trapezoid rule.

23. $|I - T_n| = |(I - L_n)/2 + (I - R_n)/2| \leq \frac{1}{2}|I - L_n| + \frac{1}{2}|I - L_n| \leq \frac{K_1 (b-a)^2}{2n}.$

7.3   Trapezoid Sums, Midpoint Sums, and the Second Derivative

1. (a) $M_2 = \frac{1}{2}(f(1/4) + f(3/4)) = \frac{e^{1/16} + e^{9/16}}{2} \approx 1.409.$

    $T_2 = \frac{1}{4}(f(0) + 2f(1/2) + f(1)) = \frac{1 + 2e^{1/4} + e^1}{4} \approx 1.571.$

    (b) In all parts below, we'll use $K_1 = 6$, $K_2 = 16.31$. Then

    $L_{10} \approx 1.381$. Error bound: $|L_{10} - I| \le \frac{K_1 \cdot 1}{20} = \frac{6}{20} = 0.3.$

    $R_{10} \approx 1.553$. Error bound: same as L_{10}.

    $M_{10} \approx 1.460$. Error bound: $|M_{10} - I| \le \frac{K_2 \cdot 1^3}{24 \cdot 100} = \frac{16.31}{2400} \approx 0.007.$

    $T_{10} \approx 1.467$. Error bound: $|T_{10} - I| \le \frac{K_2 \cdot 1^3}{12 \cdot 100} = \frac{16.31}{1200} \approx 0.0136.$

    L_{10} and M_{10} *under*estimate the exact value; the others underestimate.

    (c) We want $|M_n - I| \le \frac{K_2(b-a)^3}{24n^2} = \frac{16.31(1)^3}{24n^2} \le 0.0005$. The last inequality holds if $n^2 \ge \frac{16.31}{24(0.0005)} \approx$

    1359.17, i.e., if $n \ge \sqrt{\frac{4000}{3}} \approx 36.87$. In particular, $n = 37$ is OK so the estimate $M_{37} \approx 1.46248$ is good to 3 decimal places.

3. Since $\left|f''(x)\right| \le 2.5$ if $0 \le x \le 5$, $\left|T_n - \int_a^b f(x)\,dx\right| \le 2.5 \cdot 5^3/12n^2$. The expression on the right is less than 0.001 if $n \ge 162$.

5. $K_2 = 0$; $I = 4$; $T_{10} = 4.0000$; $|I - T_{10}| = 0$; $\frac{K_2(b-a)^3}{12n^2} = 0$; $M_{10} = 4.0000$; $|I - M_{10}| = 0$; $\frac{K_2(b-a)^3}{24n^2} = 0$

7. $K_2 = 0.25$; $I = 4.6667$; $T_{10} = 4.6648$; $|I - T_{10}| = 0.00187$; $\frac{K_2(b-a)^3}{12n^2} = 0.0056$; $M_{10} = 4.6676$;

    $|I - M_{10}| = 0.0009$; $\frac{K_2(b-a)^3}{24n^2} = 0.0028$

9. $K_2 = 1$; $I = 0.57384$; $T_{10} = 0.57337$; $|I - T_{10}| = 0.00047$; $\frac{K_2(b-a)^3}{12n^2} = 0.00083$; $M_{10} = 0.57408$;

    $|I - M_{10}| = 0.00024$; $\frac{K_2(b-a)^3}{24n^2} = 0.00042$

11. For $\int_0^2 \sin(x^2)\,dx$, $K_2 = 10.80156$ works. Therefore, $n^2 \ge \frac{10.80156 \cdot 2^3}{24 \cdot 0.005} = 720.104 \implies n \ge 27$ will do.

13. For $\int_1^{10} \sin(1/x)\,dx$, $K_2 = 0.425$ works. Therefore, $n^2 \ge \frac{0.425 \cdot 9^3}{24 \cdot 0.005} = 2581.875 \implies n \ge 51$ will do.

15. Because the function is monotone increasing and concave down, L_{30} and T_{30} underestimate I; M_{30} and R_{30} overestimate I. Thus we have $L_{30} < T_{30} < I < M_{30} < R_{30}$.

17. The graph shows that f'' is positive—hence f is *concave up*—over the interval of integration.

    (a) $L_{100} \le R_{100}$ could be true or false, depending on whether f is increasing or decreasing. The graph of f'' doesn't tell which is the case.

    (b) $T_{200} \le M_{200}$ is false: If the graph of f is concave up, T_n must *overestimate* I and M_n must *underestimate*. Thus $T_n > M_n$ for any n.

    (c) $M_{50} \le L_{50}$ could be true or false, depending on whether f is increasing or decreasing. The graph of f'' doesn't tell which is the case.

19. (a) **Must** be true. Since f is concave upwards on the interval of integration, any trapezoidal rule estimate overestimates I (i.e., $I - T_n < 0$).

    (b) **Cannot** be true. Since f is decreasing on the interval of integration $M_n > R_n$ for every n. [If f is decreasing on an interval $[a, b]$, then $m = (a + b)/2$ is the midpoint of the interval and $f(m) > f(b)$.]

    (c) **Cannot** be true. Because f is decreasing and concave upwards on the interval of integration, both L_n and T_n overestimate I for any n. Also, $T_n < L_n$ since $R_n < L_n$ (f is decreasing) and $T_n = (L_n + R_n)/2$. Together, these results imply that $L_n - I > T_n - I > 0 \iff |I - L_n| > |I - T_n|$.

21. If $f(x) = x^2$, then f is concave up, so M_n underestimates I and the approximation error is as bad as Theorem 3 allows.

23. If $f(x) = -x^2$, then f is concave down, so T_n underestimates I and the approximation error is as bad as Theorem 3 allows.

25. (a) The error bounds for M_n and M_{10n} are

$$|I - M_n| \le \frac{K_2(b - a)^3}{24n^2}; \qquad |I - M_{10n}| \le \frac{K_2(b - a)^3}{24 \cdot (10n)^2} = \frac{K_2(b - a)^3}{24 \cdot 100n^2}.$$

The second bound has an extra factor of 100 in the denominator. This means that using ten times as many subintervals in M_n gives about *two* extra decimal places of accuracy.

    (b) The error bounds for L_n and L_{10n} are

$$|I - L_n| \le \frac{K_1(b - a)^2}{2n}; \qquad |I - L_{10n}| \le \frac{K_1(b - a)^2}{2 \cdot 10n} = \frac{K_1(b - a)^2}{20n}.$$

Here the second bound has an extra factor of 10 in the denominator, using ten times as many subintervals in L_n gives only *one* extra decimal place of accuracy.

27. (a) Here's a table of values for the various integrals. (The final column is sensitive to roundoff errors—so results may vary.)

Exercise	I	T_{10}	M_{10}	$I - T_{10}$	$I - M_{10}$	$\dfrac{I - M_{10}}{I - T_{10}}$
5. $\int_1^2 x^2\, dx$	2.3333	2.3350	2.3325	-0.00167	0.00083	-0.5000
6. $\int_1^4 \sqrt{x}\, dx$	4.6667	4.6648	4.6676	0.001871	-0.000934	-0.4992
7. $\int_1^2 x^{-1}\, dx$	0.6931	0.6938	0.6928	-0.000624	0.000312	-0.4995
8. $\int_2^3 \sin x\, dx$	0.5738	0.5734	0.5741	0.000478	-0.000239	-0.5001

The results show the general pattern: the T_{10} error is about *twice* the M_{10} error, and the errors are *opposite in sign*.

    (b) $I = \int_0^1 \sqrt{x}\, dx = 2/3$; $T_{10} \approx 0.6605$; $M_{10} \approx 0.6684$; $I - T_{10} \approx 0.006157$, $I - M_{10} \approx -0.001717$, $(I - M_{10})/(I - T_{10}) \approx -0.27889$.

The surprise here is that this time the last quantity is not near -0.5 as in each case in the previous part. The difference is that the graph of this integrand is vertical at the left endpoint of integration. This fouls up the usual error estimates.

7.4 Simpson's Rule

1. For $I = \int_a^b dx = x\big]_a^b = b - a$. We need to show that S_2 has the *same* value as I. Here goes:

$$S_2 = \frac{b-a}{6}\big(f(a) + 4f((a+b)/2) + f(b)\big) = \frac{b-a}{6}(1 + 4 \cdot 1 + 1) = b - a.$$

3. For $I = \int_a^b x^2\, dx = \frac{x^3}{3}\Big]_a^b = \frac{b^3 - a^3}{3}$. We need to show that S_2 has the *same* value as I. Here goes:

$$S_2 = \frac{b-a}{6}\big(f(a) + 4f((a+b)/2) + f(b)\big) = \frac{b-a}{6}\left(a^2 + 4\frac{(a+b)^2}{4} + b^2\right)$$

$$= \frac{b-a}{6}\left(2a^2 + 2ab + 2b^2\right) = \frac{b-a}{3}\left(a^2 + ab + b^2\right) = \frac{b^3 - a^3}{3}.$$

5. Let $I = \int_a^b x^4\, dx = \frac{b^5}{5} - \frac{a^5}{5}$. By comparison, $S_2 = \frac{b-a}{6}\left(a^4 + 4((b+a)/2)^4 + b^4\right)$.

Some careful algebra now shows: $|I - S_2| = \left|\frac{b^5 - a^5}{5} - \frac{(b-a)}{6}\left(a^4 + \frac{(b+a)^4}{4} + b^4\right)\right| = \frac{(b-a)^5}{120}$.

The error bound is: $|S_2 - I| \le \frac{K_4(b-a)^5}{180n^4} = \frac{24(b-a)^5}{180 \cdot 2^4} = \frac{(b-a)^5}{120}$. Thus S_2 does commit the maximum possible error.

7. (a) $I = \int_0^1 \cos(100x)\, dx = \frac{\sin(100x)}{100}\Big]_0^1 = \frac{\sin 100}{100} \approx -0.0050637$.

   (b) $S_{10} \approx 0.036019$; thus the actual approximation error is $|I - S_{10}| \approx 0.041083 < 0.05$.

   (c) From the error bound formula, with $K_4 = 100^4 = 10^8$ and $n = 10$, we get

   $$|\text{error}| \le \frac{K_4 \cdot 1^5}{180 \cdot 10^8} = \frac{10^8}{180 \cdot 10^4} \approx 55.556.$$

   This bound is large because K_4 is so large.

9. Let $I = \int_1^7 f(x)\, dx$. Since f is positive on the interval of integration, all five estimates are positive numbers. Since f is increasing on the interval of integration, $L_{100} \le I \le R_{100}$. Since f is concave down on the interval of integration, $T_{100} \le I \le M_{100}$. Furthermore, since T_{100} is the average of L_{100} and R_{100}, $L_{100} \le T_{100} \le R_{100}$. Also, S_{100} is a weighted average of T_{100} and M_{100}, so $T_{100} \le S_{100} \le M_{100}$. Therefore,

$$L_{100} \le T_{100} \le S_{100} \le M_{100} \le R_{100}.$$

11. $S_4 = \frac{1}{3}(2M_2 + T_2) = 141.425$.

   When $-1 \le x \le 2$, $\left|f^{(4)}(x)\right| \le 8 = K_4$, so $|I - S_4| \le \frac{8 \cdot 2^5}{180 \cdot 4^4} = 0.0421875$.

13. For $I = \int_0^1 \sin(\sin(x))\, dx$ we can use $K_2 = 1$ and $K_4 = 3.8$.

   (a) For the midpoint rule we have M_n error $\le \frac{1 \cdot 1}{24n^2} = \frac{1}{24n^2}$. Setting $n = 4$, $n = 8$, and $n = 16$ gives, respectively: M_4 error $\le \frac{1}{384} \approx 0.0026$; M_8 error $\le \frac{1}{4 \cdot 384} \approx 0.00065$; M_{16} error $\le \frac{1}{16 \cdot 384} \approx 0.000163$.

(b) For Simpson's rule we have S_n error $\leq \dfrac{1 \cdot 3.8}{180n^4} = \dfrac{3.8}{180n^4}$. Setting $n = 4$, $n = 8$, and $n = 16$ gives, respectively: S_4 error $\leq \dfrac{3.8}{180 \cdot 4^4} \approx 8.2465 \times 10^{-5}$; S_8 error $\leq \dfrac{3.8}{180 \cdot 8^4} \approx 5.1541 \times 10^{-6}$; S_{16} error $\leq \dfrac{3.8}{180 \cdot 16^4} \approx 3.2213 \times 10^{-7}$.

15. (a) We're given that $f^{(4)}(x) = \left(\cos^4 x - 6\cos^2 x \sin x - 4\cos^2 x + 3\sin^2 x + \sin x\right) \cdot e^{\sin x}$. Recall that (1) sines and cosines always lie in $[-1, 1]$; and (2) the absolute value of a sum is no more than the sum of the absolute values. It follows that for x in $[0, 1]$,

$$|f^{(4)}(x)| \leq (1 + 6 + 4 + 3 + 1)e^1 = 15e \approx 40.8 < 41.$$

(b) The graph of $f^{(4)}$ on $[0, 1]$ gives a better bound—it shows that $|f^{(4)}| < 11$.

(c) Using $K_4 = 11$, we see that get $|S_n - I| \leq 0.001$ if $\dfrac{11 \cdot (200\pi)^5}{180n^4} \leq 0.001$. Solving this inequality for n gives $n \geq 8795$; thus $n = 8796$ is OK. (This is a huge number—somewhat impractical for real calculations.)

(d) $I = 100 \int_0^{2\pi} f(x)\,dx$ because f is *periodic*; it *repeats* itself every 2π. On $[-50\pi, 150\pi]$, it repeats itself 100 times.

(e) For $\int_0^{2\pi} f(x)\,dx$, the Simpson error formula says error $\leq \dfrac{K_4 \cdot (2\pi)^5}{180n^4} = \dfrac{11 \cdot (2\pi)^5}{180n^4} \leq 0.00001$; solving for n gives $n \geq 87.9$. Thus $n = 88$ will do. (In fact, $S_{88} \approx 7.9549265$; this is guaranteed correct within 0.00001.)

(f) To find the original integral, we can multiply our estimate for $\int_0^{2\pi} f(x)\,dx$ (found with S_{88}) by 100; the error won't be more than $100 \cdot 0.00001 = 0.001$. (The result: $\int_{-50\pi}^{150\pi} f(x)\,dx \approx 795.493$ is correct to within 0.001.)

8.1   Introduction

1. We know that $r_B(t) = A + B(t - 35) + C(t - 35)^2$ for some constants A, B, and C. We'll show that the only possible values for A, B, and C are $A = 13$, $B = 0$, and $C = -1/100$.

   Note first that

   $$r_B(t) = A + B(t - 35) + C(t - 35)^2 \implies r_B'(t) = B + 2C(t - 35).$$

   From the problem we know that $r_B(35) = 13$, $r_B(60) = 27/4$, and $r_B'(35) = 0$. This leads to three simple conditions on A, B, and C:

   $$r_B(35) = A = 13; \quad r_B'(35) = B = 0; \quad r_B(60) = A + B \cdot 25 + C \cdot 25^2 = 27/4.$$

   Thus $A = 13$, $B = 0$, and (from the last equation) $C = -1/100$.

3. (a) The appropriate integral is $\displaystyle\int_5^{55} r_B(t)\,dt = \int_5^{55}\left(13 - \frac{(t - 35)^2}{100}\right)dt = \frac{1600}{3}$. Thus Brown harvests about 533.33 bushels from $t = 5$ to $t = 55$.

   (b) We want to estimate the integral $\int_5^{55} r_J(t)\,dt$ using T_5, the trapezoid rule with 5 subdivisions, each of length 10. Here's the result:

   $$\left(\frac{r_J(5) + r_J(15)}{2} + \frac{r_J(15) + r_J(25)}{2} + \cdots + \frac{r_J(35) + r_J(45)}{2}\right)\cdot 10 = 525.$$

   By this estimate, therefore, Jones harvests 525 bushels over the period.

5. (a) The linear function $f(x) = bx/a$ does the job. The length of its graph from $x = 0$ to $x = a$ is given by the integral

   $$\int_0^a \sqrt{1 + f'(x)^2}\,dx = \int_0^a \sqrt{1 + b^2/a^2}\,dx = a\sqrt{1 + b^2/a^2} = \sqrt{a^2 + b^2}.$$

   (An easier way to find the answer is to use the distance formula in the plane.)

   (b) If $f(x) = x^2 + 1$, then $f'(x) = 2x$, so we want the integral $I = \displaystyle\int_0^1 \sqrt{1 + f'(x)^2}\,dx = \int_0^1 \sqrt{1 + 4x^2}\,dx$. We estimated this same integral in Example 2, where we got $M_{20} \approx 1.479$.

   (c) For each of the two curves in question we have $dy/dx = \cos x$, so the length integral is the same in each case: $I = \displaystyle\int_0^\pi \sqrt{1 + f'(x)^2}\,dx = \int_0^\pi \sqrt{1 + \cos^2 x}\,dx$. For this integral, $M_{20} \approx 3.820$—that's a good estimate for the length of both curves.

   (d) If $g(x) = f(x) + C$, for any constant C, then $g'(x) = f'(x)$, so the length of the g-graph from $x = a$ to $x = b$ is $\displaystyle\int_a^b \sqrt{1 + g'(x)^2}\,dx = \int_a^b \sqrt{1 + f'(x)^2}\,dx$. The last quantity is independent of C.

   Geometrically, the idea is that adding a constant C to f raises or lowers the graph of f, but *doesn't change its length*.

7. (a) $I = \displaystyle\int_0^1 \sqrt{1 + x}\,dx = \frac{2}{3}(1 + x)^{3/2}\Big]_0^1 = \frac{4\sqrt{2} - 2}{3} \approx 1.219.$

   (b) I is the area under the curve $y = \sqrt{1 + x}$ from $x = 0$ to $x = 1$. (The region in question is more or less trapezoidal, with base 1 and altitudes 1 and $\sqrt{2}$.)

   (c) We want a function f for which

   $$I = \int_0^1 \sqrt{1 + x}\,dx = \int_0^1 \sqrt{1 + f'(x)^2}\,dx.$$

Let's look, therefore, for an f for which $f'(x)^2 = x$, or $f'(x) = \sqrt{x}$. *Any* antiderivative of $\sqrt{x}$ will do; let's use $f(x) = 2x^{3/2}/3$. Plotting this f over $[0, 1]$ gives a graph whose length appears to be around 1.2, as the previous part suggests.

(d) Any antiderivative of $\sqrt{x}$, i.e., any function of the form $f(x) = \frac{2}{3}x^{3/2} + C$, where C is a constant. (There are other possibilities, too. For instance, we could use $f(x) = -2x^{3/2}/3$, the *opposite* of the function in the previous part.

8.2   Finding Volumes by Integration

1. $V = \int_0^8 \pi \left(x^3\right)^2 dx = \dfrac{8^7 \pi}{7}$

3. $V = \int_0^2 \pi \left((x+6)^2 - \left(x^3\right)^2\right) dx = \dfrac{1688\pi}{21}$

5. $V = \int_0^2 \pi \left(4^2 - \left(y^2\right)^2\right) dy = \dfrac{128\pi}{5}$

7. $V = \int_0^4 \pi \left(\sqrt{y}\right)^2 dy - \int_1^4 \pi \left(\log_2 y\right)^2 dy = \pi \left(\dfrac{16}{\ln 2} - 8 - \dfrac{6}{(\ln 2)^2}\right) \approx 8.15214$

9. At height y above the base, a cross section parallel to the base of the cone is a circle of radius $\frac{r}{h}(h-y)$. Thus, the volume of the cone is

$$V = \int_0^h \pi \left(\frac{r}{h}(h-y)\right)^2 dy = \frac{\pi r^2}{h^2} \int_{-h}^0 u^2 \, du = \frac{\pi r^2 h}{3}.$$

[The substitution $u = h - y$ was used to evaluate the integral.]

Alternatively, if the vertex of the cone is placed at the origin and the center of the base is placed on the positive x-axis, the area of a cross-section parallel to the base is $A(x) = \pi (rx/h)^2$. Thus, the volume of the cone is

$V = \dfrac{\pi r^2}{h^2} \int_0^h x^2 \, dx = \dfrac{\pi r^2 h}{3}.$

11. (a) $\pi \int_0^4 ((6-y) - (-2))^2 \, dy - \pi \int_0^4 (\sqrt{y} - (-2))^2 \, dy$

   (b) $\pi \int_0^2 (x^2 - (-1))^2 \, dx + \pi \int_2^6 ((6-x) - (-1))^2 \, dx$

13. $V = \int_{-1}^2 \pi \left(((x+1)+1)^2 - \left((x^2-1)+1\right)^2\right) dx = \dfrac{72\pi}{5}$

15. $V = \int_0^1 \pi \left((\sqrt{x}+2)^2 - \left(x^2+2\right)^2\right) dx = \dfrac{49\pi}{30}$

17. A circle of circumference c has area $\frac{c^2}{4\pi}$. Thus, since 1 foot = 12 inches, the volume of the pole is approximately

$$S_6 = \frac{60 \cdot 12}{3 \cdot 6 \cdot 4\pi} \left(1 \cdot 16^2 + 4 \cdot 14^2 + 2 \cdot 10^2 + 4 \cdot 5^2 + 2 \cdot 3^2 + 4 \cdot 2^2 + 1 \cdot 1^2\right) = \frac{13750}{\pi} \approx 4376.8 \text{ in}^3.$$

19. The radius of the glass (in inches) at height h (in inches) is $r = 1 + 0.1h$ when $0 \le h \le 5$. Thus, the volume of the glass is $V = \pi \int_0^5 (1+0.1h)^2 \, dh = \dfrac{95\pi}{12} \approx 24.87 \text{ in}^3$.

21. The volume of water in the balloon is $V = \pi \int_{-3}^1 (9 - y^2) \, dy = 80\pi/3$ cubic inches.

23. (a) $V = \pi \int_{-3}^{-1} \arctan^2 x \, dx$

   (b) For any n, L_n overestimates V since $\left(\arctan^2 x\right)' = \dfrac{2 \arctan x}{1 + x^2} < 0$ over the interval of integration (i.e., the integrand is a decreasing function).

25. Cross-sections parallel to the ends of the tank have area:

$$A(x) = 2 \int_{-6}^{3} \sqrt{9 - \frac{y^2}{4}} \, dy = 4 \int_{-3}^{3/2} \sqrt{9 - u^2} \, du = 2 \left(u\sqrt{9 - u^2} + 9 \arcsin \left(\frac{u}{3} \right) \right) \Big|_{-3}^{3/2} = \frac{9\sqrt{3}}{2} + 12\pi.$$

Thus, the volume of fuel oil in the tank is $V = \int_{0}^{10} A(x) \, dx = 45\sqrt{3} + 120\pi \approx 454.93$ cubic feet.

27. (a) The area of the annulus is $\pi(r + \Delta r)^2 - \pi r^2 = \pi(2r + \Delta r)\Delta r$.

    (b) The area of a circle of radius R is is the area of a circle of radius r and $n - 1$ concentric annuli with thickness Δr. The area of the circle of radius r is $\pi(\Delta r)^2$ and the area of the kth annulus is $\pi(2r_k + \Delta r)\Delta r$.

    Therefore, since $r_0 = 0$, the area of the circle of radius R is $\sum_{k=0}^{n-1} \pi(2r_k + \Delta r)\Delta r = \pi R^2$.

    (c) Observe that $\lim_{n \to \infty} \sum_{k=0}^{n-1} 2\pi r_k \Delta r = \int_{0}^{R} 2\pi r \, dr$ — the sum on the left is a left sum approximation to the

    integral on the right. Therefore, since $\lim_{n \to \infty} \sum_{k=0}^{n-1} (\Delta r)^2 = \lim_{n \to \infty} \sum_{k=0}^{n-1} \left(\frac{R}{n} \right)^2 = \lim_{n \to \infty} \frac{R^2}{n} = 0$.

    $$\lim_{n \to \infty} \sum_{k=0}^{n-1} \pi(2r_k + \Delta r)\Delta r = \int_{0}^{R} 2\pi r \, dr.$$

    (d) This is an alternate algebraic representation of the sum in part (b). The area of a circle of radius R is written as the sum of the area of a circle and the areas of $n - 1$ annuli, each with thickness Δr.

29. (a) Since f is a decreasing function, each left cylindrical shell overestimates the volume it encloses. Similarly, each right shell underestimates the volume it encloses. Therefore, the left and right sums overestimate and underestimate, respectively, the volume V.

    (b) $\lim_{n \to \infty} \sum_{k=1}^{n} \pi(2x_k - \Delta x) f(x_k)\Delta x = \lim_{n \to \infty} \sum_{k=0}^{n-1} \pi(2x_k + \Delta x) f(x_k)\Delta x = \int_{a}^{b} 2\pi x f(x) \, dx = V$.

    (c) The roles of the left and right sums in parts (a) and (b) are interchanged when f is an increasing function. Otherwise, the ideas are the same.

31. (a) The racetrack principle implies that $f(z) \geq f(c) - K(z - c)$. Since $d \geq z$, it follows that $f(z) \geq f(c) - K(z - d)$. The result that $f(z) \leq f(c) + K(d - c)$ can be derived using a similar argument.

    (b) Using part (a), we find that $f(x_k) - K\Delta x \leq f(z) \leq f(x_k) + K\Delta x$ when $x_k \leq z \leq x_{k+1}$. Thus, $\pi(2x_k - \Delta x)(f(x_k) - K\Delta x)\Delta x$ underestimates the volume of the "shell" obtained by rotating the area under the graph of f between x_k and x_{k+1}. Similarly, $\pi(2x_k + \Delta x)(f(x_k) + K\Delta x)\Delta x$ overestimates the volume of the "shell" obtained by rotating the area under the graph of f between x_k and x_{k+1}.

    (c) Since $\lim_{n \to \infty} (\Delta x)^2 = 0$, $\lim_{n \to \infty} (\Delta x)^3 = 0$, and f is bounded on the interval $[a, b]$,

    $$\lim_{n \to \infty} \sum_{k=0}^{n-1} \pi(2x_k - \Delta x)(f(x_k) - K\Delta x)\Delta x = \lim_{n \to \infty} \sum_{k=0}^{n-1} \pi(2x_k + \Delta x)(f(x_k) + K\Delta x)\Delta x = \int_{a}^{b} 2\pi x f(x) \, dx.$$

33. (a) Solving the equation $y = x\sqrt{1 - x^2}$ for x^2 yields $x^2 = \left(1 \pm \sqrt{1 + 4y^2} \right) / 2$. Thus, the outer boundary of the region is the curve $x = \left(1 + \sqrt{1 + 4y^2} \right)^{1/2}$ and the inner boundary of the region is the curve $x = \left(1 - \sqrt{1 + 4y^2} \right)^{1/2}$ when $0 \leq y \leq 1/2$. Therefore, the volume of the solid of revolution is

    $$V = \int_{0}^{1/2} \frac{\pi}{2} \left(1 + \sqrt{1 - 4y^2} \right) dy - \int_{0}^{1/2} \frac{\pi}{2} \left(1 - \sqrt{1 - 4y^2} \right) dy.$$

(b) Using the method of cylindrical shells, $V = \int_0^1 \pi x f(x)\, dx = \int_0^1 2\pi x^2 \sqrt{1-x^2}\, dx$.

(c) The integral in part (b) can be evaluated using the substitutions $u = 1 - x^2$, $w = u - 1/2$ and the table of integrals in the back of the textbook:

$$\int_0^1 2\pi x^2 \sqrt{1-x^2}\, dx = \pi \int_0^1 \sqrt{(1/2)^2 - (u - 1/2)^2}\, du = \pi \int_{-1/2}^{1/2} \sqrt{(1/2)^2 - w^2}\, dw = \frac{\pi^2}{8}.$$

35. (a) Let $p(r)$ be the population density at a distance r from the center of the city. Since p is a linear function it can be written in the form $p(r) = ar + b$. Now, $p(0) = K$ and $p(R) = 0$, so $b = K$ and $a = -K/R$. Thus, $p(r) = -Kr/R + K = K(1 - r/R)$.

(b) The population of the city is $\int_0^R p(r)\, dr = \dfrac{KR}{2}$.

8.3   Arclength

1. The integral is $\int_0^1 \sqrt{1+1}\,dx = \sqrt{2}$. This result is (of course) the same as that given by the usual distance formula.

3. (a) Here's a picture. Since the line connecting the endpoints of the curve C has length
   $\sqrt{(3-1)^2 + (109/12 - 7/12)^2} = \sqrt{305/4} \approx 8.7321$, the length of the curve is slightly more than this.

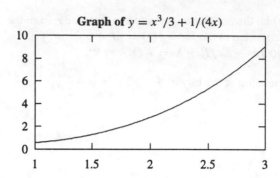

Graph of $y = x^3/3 + 1/(4x)$

(b) By the arclength formula, the length is the integral:

$$\int_1^3 \sqrt{1+(f')^2} = \int_1^3 \sqrt{1 + \left(x^2 - \frac{1}{4x^2}\right)^2}\,dx = \int_1^3 \sqrt{1 + x^4 - \frac{1}{2} + \frac{1}{16x^4}}\,dx$$

$$= \int_1^3 \sqrt{x^4 + \frac{1}{2} + \frac{1}{16x^4}}\,dx = \int_1^3 \sqrt{\left(x^2 + \frac{1}{4x^2}\right)^2}\,dx$$

$$= \int_1^3 \left(x^2 + \frac{1}{4x^2}\right)\,dx = \frac{x^3}{3} - \frac{1}{4x}\Bigg]_1^3$$

$$= \frac{53}{6} \approx 8.833333333.$$

5. The shortest distance between two points A and B is a line. If the curve passing through the points A and B is not a line, then any polygonal path from A to B through intermediate points on the curve will be longer than the line segment connecting the two points.

7. length $= \int_0^1 \sqrt{1 + e^{2x}}\,dx = \frac{1}{2}\int_1^{e^2} \frac{\sqrt{1+u}}{u} = \frac{1}{2}\int_1^{e^2} \frac{1+u}{u\sqrt{1+u}} = \left(\sqrt{1+u} + \frac{1}{2}\ln\left|\frac{\sqrt{1+u}-1}{\sqrt{1+u}-1}\right|\right)\Bigg|_1^{e^2}$

   $\approx 2.003497110.$

9. (a) The length of the curve $y = f(x)$ from $x = a$ to $x = b$ is $\int_a^b \ell(x)\,dx$, where $\ell(x) = \sqrt{1 + (f'(x))^2}$. Now,
   $\ell'(x) = f'(x)f''(x)/\sqrt{1 + (f'(x))^2}$ so $\ell'(x) < 0$ on interval $[a, b]$ because $f'(x) > 0$ and $f''(x) < 0$ when $a \le x \le b$. Thus, $\ell(x)$ is a decreasing function on the interval $[a, b]$ and, therefore, L_n *over*estimates the value of the arclength integral.

   (b) Let $\ell(x) = \sqrt{1 + (g'(x))^2}$. Then, $\left|L_{10} - \int_0^1 \ell(x)\,dx\right| \le \frac{K_1}{20}$, where K_1 is a number such that $|\ell'(x)| \le K_1$ when $0 \le x \le 1$. Since $4 \le g'(x) \le 7$ and $-3 \le g''(x) \le -2$ when $0 \le x \le 1$,

   $$|\ell'(x)| = \left|\frac{g'(x)g''(x)}{\sqrt{1 + (g'(x))^2}}\right| \le \left|\frac{7 \cdot -3}{\sqrt{1 + (4)^2}}\right| = \frac{21}{\sqrt{17}}.$$

Therefore, the left sum estimate L_{10} approximates the arclength integral within $21/20\sqrt{17} \approx 0.25466$.

11. After 5 minutes, the bug has crawled 30 feet. Since the length of the curve $y = \frac{1}{3}\left(x^2 + 2\right)^{3/2}$ between $x = 1$ and $x = s$ is $s^3/3 + s - 4/3$, $x \approx 4.3271$ is the x-coordinate of the bug's position after 5 minutes. Thus, the bug is (approximately) at the point $(4.3271, 31.4470)$.

8.4 Work

1. (a) work $= 115 \times 0.06$ inch-pounds $= 6.9$ inch-pounds $= 0.575$ foot-pounds.

   (b) According to Hooke's Law, the force required to compress a spring x units is proportional to x. Since a force of 115 pounds compresses the spring by 0.06 inches, the constant of proportionality (the *spring constant* is $k = 115/0.06$. Therefore, a force of 175 pounds will compress the spring approximately 0.0913 inches and work $= 175 \times 0.0913$ inch-pounds $= 15.978$ inch-pounds $= 1.3315$ foot-pounds.

3. Each parallel slice is a circular cylinder with cross-sectional area 25π and thickness Δx. Therefore,

$$\text{work} = \int_0^5 62.4 \cdot 25\pi \cdot (10 - x)\, dx = 58{,}500\pi \text{ foot-pounds} \approx 183{,}783 \text{ foot-pounds.}$$

5. Let x denote the number of inches of compression.

   (a) To find k, use the equation (implicit in the problem statement) $F(2) = 2k = 10$. It follows that $k = 5$ (and hence that $F(x) = 5x$.)

   (b) Compressing from 16 inches to 12 inches means compressing from $x = 2$ to $x = 6$. Thus the work done is

$$W = \int_2^6 F(x)\, dx = \int_2^6 5x\, dx = 80 \text{ inch-pounds.}$$

7. work $= \displaystyle\int_{-4}^4 42 \cdot 2\sqrt{16 - y^2} \cdot 15 \cdot (y + 17)\, dy$

   $= \left. -420\left(16 - y^2\right)^{3/2} + 10{,}710 y\sqrt{16 - y^2} + 171{,}360 \arcsin(y/4) \right|_{-4}^{4}$

   $= 171{,}360\pi \text{ foot-pounds} \approx 538{,}343 \text{ foot-pounds}$

9. (a) $W = \displaystyle\int_0^2 F(x)\, dx = \int_0^2 40x\, dx = 80$ foot-pounds.

   (b) $W = \displaystyle\int_0^s F(x)\, dx = \int_0^s 40x\, dx = 20s^2$ foot-pounds.

   (c) By the problem and the previous part, $W = 20s^2 = 10000$. Solving this for s gives $s = 10\sqrt{5} \approx 22.3361$ feet.

11. Let x denote the distance (in feet) that the spring is extended. Notice that since the chain weighs 20 pounds, we start with $x = 5$. (Draw a picture! Note, too, that other x-scales are possible.)

   The problem is to find $\int_5^7 F(x)\, dx$, where $F(x)$ is the net downward force necessary at a given x.

   Let's find a formula for $F(x)$; there are two main ingredients: (i) For any value of x, the spring exerts an *upward* force of $4x$ pounds. (ii) For a given value of x, the length of chain remaining above the floor is $10 - (x - 5) = 15 - x$ feet. (A diagram should make this convincing.) Since the chain weighs 2 pounds per foot, this length of chain exerts a *downward* force of $2(15 - x) = 30 - 2x$ pounds.

   Putting (i) and (ii) together means that the *net* downward force required for given x is $F(x) = 4x - (30 - 2x) = 6x - 30$ pounds. Thus the desired work is

$$W = \int_5^7 F(x)\, dx = \int_5^7 (6x - 30)\, dx = 12 \text{ foot-pounds.}$$

8.5 Present Value

1. For any interest rate r, the present value of one $1 million payment, 23 years ahead, is $PV = 1,000,000 \cdot e^{-r \cdot 23}$.

   (a) If $r = 0.06$, then $PV = \$1,000,000 e^{-0.06 \cdot 23} \approx \$251,579$.

   (b) If $r = 0.08$, then $PV = \$1,000,000 e^{-0.08 \cdot 23} \approx \$155,817$.

   (c) To find the desired r we solve the equation $PV = 100,000 = 1,000,000 e^{-r \cdot 23}$ for r. The result: $r = (\ln 10)/23 \approx 0.10011$—just a bit above 10%.

3. For any interest rate r (real or nominal) the present value of one $1 million payment, 23 years ahead, is $PV = 1,000,000 \cdot e^{-r \cdot 23}$.

   (a) If $r = 0.02$, then $PV = \$1,000,000 e^{-0.02 \cdot 23} \approx \$631,284$.

   (b) If $r = 0.04$, then $PV = \$1,000,000 e^{-0.04 \cdot 23} \approx \$398,519$.

   (c) To find Betty's r we solve the equation $PV = 200,000 = 1,000,000 e^{-r \cdot 23}$ for r. The result: $r = (\ln 5)/23 \approx 0.07$. Thus Betty needs to find a *real* interest rate—after inflation—of 7%.

5. At any interest rate r, the present value formula for several future payments says, in this situation, that

$$PV = 40,000 e^{-18r} + 42,000 e^{-19r} + 44,000 e^{-20r} + 46,000 e^{-21r}.$$

   If $r = 0.06$, the formula (and some electronic help) give $PV \approx \$53,316.85$. If $r = 0.08$, $PV \approx \$36,119.66$.

7. (a) p_I is the graph which peaks at $t = 180$. When $t = 180$, $p_I = 110$.

   (b) The function $\cos t$ has period 2π. The functions p_T and p_I both have period 360.

   (c) The constant 50 affects the amplitude of the graphs. The constant 60 shifts the graphs upward. The constants 180, 105, and π affect where each graph has its local maxima and minima.

9. The total return from the investment is $40,000 paid continuously over an 8-year time interval. Since the present value of the return from the investment is $PV = \$5,000 \int_0^8 e^{-0.06t} \, dt \approx \$31,768$, this is a worthwhile investment.

11. (a) $\left(\dfrac{e^{ax}}{a^2 + b^2} (a \cos(bx) + b \sin(bx)) \right)' = \dfrac{a e^{ax}}{a^2 + b^2} (a \cos(bx) +$

$$b \sin(bx)) + \frac{e^{ax}}{a^2 + b^2} (-ab \sin(bx) + b^2 \cos(bx))$$

$$= e^{ax} \cos(bx)$$

   (b) $\displaystyle\int_0^{360} \left(50 \cos \left(\pi \cdot \frac{t - 180}{180} \right) + 60 \right) e^{-0.1t/360} \, dt = \left(216,000 - \frac{180,000}{1 + 400\pi^2} \right) \left(1 - e^{-1/10} \right) \approx 20,550.78$

   NOTE: $\cos \left(\pi \cdot \dfrac{t - 180}{180} \right) = \cos \left(\dfrac{\pi t}{180} - \pi \right) = -\cos \left(\dfrac{\pi t}{180} \right)$

13. The interest earned from the income received at time t is $p(t) e^{r(T-t)}$. Thus, the total income accrued between $t = 0$ and $t = T$ is $\displaystyle\int_0^T p(t) e^{r(T-t)} \, dt = e^{rT} \int_0^T p(t) e^{-rt} \, dt = e^{rT} \, PV$.

8.6   Fourier Polynomials

1. (a) $f(x) = f'(x) = f''(x) = f'''(x) = e^x$ so $f(0) = f'(0) = f''(0) = f'''(0) = e^0 = 1$.

    Now, $p(0) = 1$; $p'(x) = 1 + x + x^2/2$ so $p'(0) = 1$; $p''(x) = 1 + x$ so $p''(0) = 1$; $p'''(x) = 1$ so $p'''(0) = 1$.

3. Graphs show that the function $\sin(kx)$ is odd, so the integral is zero. For $\cos(kx)$, the symmetry about $x = 0$ gives the same result. The other part is a straightforward symbolic calculation.

5. All summands give zero integrals except the constant term.

7. The integrand is odd because it's the product of an odd function and an even function. Integrating any odd function over $[-\pi, \pi]$ gives zero.

9. (b) $a_0 = \dfrac{1}{2\pi} \displaystyle\int_{-\pi}^{\pi} f(x)\,dx = \dfrac{1}{2\pi} \int_{0}^{\pi} dx = \dfrac{1}{2}$

    $a_k = \dfrac{1}{\pi} \displaystyle\int_{-\pi}^{\pi} f(x) \cos(kx)\,dx = \dfrac{1}{\pi} \int_{0}^{\pi} \cos(kx)\,dx = \dfrac{\sin(kx)}{k\pi}\Big]_{0}^{\pi} = 0$

    $b_k = \dfrac{1}{\pi} \displaystyle\int_{-\pi}^{\pi} f(x) \sin(kx)\,dx = \dfrac{1}{\pi} \int_{0}^{\pi} \sin(kx)\,dx = -\dfrac{\cos(kx)}{k\pi}\Big]_{0}^{\pi} = \dfrac{1 - \cos(k\pi)}{k\pi}.$

    Thus, $b_{2m} = 0$ and $b_{2m+1} = \dfrac{2}{(2m + 1)\pi}$ for $m = 1, 2, 3, \ldots$.

    (c) $q_1(x) = \dfrac{1}{2} + \dfrac{2\sin x}{\pi}$;

    $q_3(x) = \dfrac{1}{2} + \dfrac{2\sin x}{\pi} + \dfrac{2\sin(3x)}{3\pi}$;

    $q_5(x) = \dfrac{1}{2} + \dfrac{2\sin x}{\pi} + \dfrac{2\sin(3x)}{3\pi} + \dfrac{2\sin(5x)}{5\pi}$;

    $q_7(x) = \dfrac{1}{2} + \dfrac{2\sin x}{\pi} + \dfrac{2\sin(3x)}{3\pi} + \dfrac{2\sin(5x)}{5\pi} + \dfrac{2\sin(7x)}{7\pi}$

9.1   Integration by Parts

1. $\int xe^{2x}\,dx = \dfrac{xe^{2x}}{2} - \dfrac{e^{2x}}{4} + C$ $[du = dx, v = e^{2x}/2]$

3. $\int x\sec^2 x\,dx = x\tan x + \ln|\cos x| + C$ $[du = dx, v = \tan x]$

5. $\int x\sqrt{1+x}\,dx = \dfrac{2}{3}x(1+x)^{3/2} - \dfrac{4}{15}(1+x)^{5/2} + C$ $[du = dx, v = \frac{2}{3}(1+x)^{3/2}]$

7. $\displaystyle\int_0^\pi x\cos(2x)\,dx = \dfrac{1}{2}x\sin(2x) + \dfrac{1}{4}\cos(2x)\bigg]_0^\pi = 0.$ $[M_2 = 0]$

9. $\displaystyle\int_1^e x\ln x\,dx = \dfrac{1}{2}x^2\ln x - \dfrac{1}{4}x^2\bigg]_1^e = \dfrac{1}{4}\left(e^2 + 1\right) \approx 2.09726.$ $[M_2 \approx 2.0670]$

11. $\displaystyle\int_{-1}^{\sqrt{2}/2} x^2\arctan x\,dx = \dfrac{1}{3}x^3\arctan x - \dfrac{1}{6}x^2 + \dfrac{1}{6}\ln\left(1+x^2\right)\bigg]_{-1}^{\sqrt{2}/2}$

$= \dfrac{\sqrt{2}}{12}\arctan(\sqrt{2}/2) + \dfrac{1}{12} + \dfrac{1}{6}\ln(3/4) - \dfrac{\pi}{12} \approx -0.15388.$ $[M_2 \approx -0.12765]$

13. Let $u = x$ and $dv = \cos^2 x\,dx = \frac{1}{2}(1 + \cos(2x))\,dx$. Then $du = dx$ and $v = \frac{1}{2}\left(x + \frac{1}{2}\sin(2x)\right)$. Therefore,

$\int x\cos^2 x\,dx = \dfrac{1}{2}x^2 + \dfrac{1}{4}x\sin(2x) - \dfrac{1}{2}\int\left(x + \dfrac{\sin(2x)}{2}\right)dx = \dfrac{1}{4}x^2 + \dfrac{1}{4}x\sin(2x) + \dfrac{1}{8}\cos(2x) + C.$

15. (a) Let $u = \sin x$ and $dv = \sin x\,dx$. Then, $du = \cos x\,dx$ and $v = -\cos x$ so

$\int \sin^2 x\,dx = -\sin x\cos x + \int\cos^2 x\,dx.$

(b) $\int \sin^2 x\,dx = -\sin x\cos x + \int\cos^2 x\,dx = -\sin x\cos x + \int\left(1 - \sin^2 x\right)dx.$

Therefore, $2\int\sin^2 x\,dx = x - \sin x\cos x$

17. $\int \arccos x\,dx = x\arccos x - \sqrt{1 - x^2} + C$ $[u = \arccos x, dv = dx]$

19. $\int (\ln x)^2\,dx = x(\ln x)^2 - 2x\ln x + 2x + C$ $[u = (\ln x)^2, dv' = dx \text{ or } u = \ln x, dv = \ln x\,dx]$

21. $\int xe^x\sin x\,dx = \dfrac{1}{2}(1 - x)e^x\cos x + \dfrac{1}{2}xe^x\sin x + C.$ $[u = x, dv = e^x\sin x\,dx]$

23. $\int x\arctan x\,dx = \dfrac{1}{2}\left(x^2\arctan x - x + \arctan x\right) + C$

Use integration by parts with $u = \arctan x$ and $dv = x\,dx$ to show that $\int x\arctan x\,dx = \frac{1}{2}\left(x^2\arctan x - \int x^2/(1+x^2)\,dx\right)$. The last antiderivative can be found using the algebraic identity

$$\dfrac{x^2}{1+x^2} = \dfrac{(1+x^2) - 1}{1+x^2} = 1 - \dfrac{1}{1+x^2}$$

25. $\int x^5\sin\left(x^3\right)dx = \dfrac{\sin\left(x^3\right)}{3} - \dfrac{x^3\cos\left(x^3\right)}{3} + C$

First substitute $w = x^3$, $dw = 3x^2\,dx$, to get the new integral $\frac{1}{3}\int w\sin w\,dw$. Now use parts, with $u = w, dv = \sin w\,dw$.

27. $\int \sqrt{x}\, e^{-\sqrt{x}}\, dx = -2e^{-\sqrt{x}}\left(x + 2\sqrt{x} + 2\right) + C$

First substitute $w = \sqrt{x}$, $w^2 = x$, $2w\, dw = dx$. This gives the new integral $2\int w^2 e^{-w}\, dw$. Finding the latter integral requires using integrations by parts twice.

29. $\int \dfrac{\arctan\left(\sqrt{x}\right)}{\sqrt{x}}\, dx = 2\sqrt{x}\arctan\left(\sqrt{x}\right) - \ln(1+x) + C$

Substitute $w = \sqrt{x}$, then use integration by parts with $u = \arctan w$ and $dv = dw$.

31. Using integration by parts with $u = \arcsin x$ and $dv = x^2\, dx$,

$$\int x^2 \arcsin x\, dx = \frac{1}{3}x^3 \arcsin x - \frac{1}{3}\int \frac{x^3}{\sqrt{1-x^2}}\, dx.$$

To find the remaining antiderivative, use the substituion $w = 1 - x^2$:

$$\int \frac{x^3}{\sqrt{1-x^2}}\, dx \rightarrow -\frac{1}{2}\int \frac{1-w}{\sqrt{w}}\, dw = -\frac{1}{2}\int \frac{dw}{\sqrt{w}} + \frac{1}{2}\int \sqrt{w}\, dw$$

$$= -\sqrt{w} + \frac{1}{3}w^{3/2} + C \rightarrow -\sqrt{1-x^2} + \frac{1}{3}(1+x^2)^{3/2} + C$$

Therefore, $\int x^2 \arcsin x\, dx = \dfrac{1}{3}x^3 \arcsin x + \dfrac{1}{3}\sqrt{1-x^2} - \dfrac{1}{9}(1+x^2)^{3/2} + C.$

33. (a) $I_1 = \int x(\ln x)^1\, dx$. Let $u = \ln x$, $dv = x\, dx$; $du = \dfrac{1}{x}\, dx$; $v = \dfrac{x^2}{2}$. Then

$$I_1 = \int x(\ln x)^1\, dx = \frac{x^2}{2}\ln x - \int \frac{x}{2}\, dx = \frac{x^2}{2}\ln x - \frac{x^2}{4} + C$$

(b) $I_n = \int x(\ln x)^n\, dx$. If we let $u = (\ln x)^n$; $dv = x\, dx$; $du = u(\ln x)^{n-1} \cdot \dfrac{1}{x}\, dx$; $v = \dfrac{x^2}{2}$. then we get

$$I_n = \frac{(\ln x)^n x^2}{2} - \frac{n}{2}\int x(\ln x)^{n-1}\, dx = \frac{x^2}{2}(\ln x)^n - \frac{n}{2}I_{n-1}$$

(c) By reduction formula plus the fact that $I_1 = \dfrac{x^2}{2}\ln x - \dfrac{x^2}{4} + C$, we have

$$I_2 = \frac{x^2}{2}(\ln x)^2 - I_1 = \frac{x^2}{2}(\ln x)^2 - \left(\frac{x^2}{2}\ln x - \frac{x^2}{4}\right) + C = \frac{x^2}{2}\left((\ln x)^2 - \ln x + \frac{1}{2}\right) + C.$$

Similarly, $I_3 = \dfrac{x^2}{2}(\ln x)^3 - \dfrac{3}{2}I_2 = \dfrac{x^2}{2}\left((\ln x)^3 - \dfrac{3}{2}(\ln x)^2 + \dfrac{3}{2}\ln x - \dfrac{3}{4}\right) + C.$

(d) This is an immediate consequence of the Fundamental Theorem of Calculus.

(e) Carrying out the differentiation on the right side of the identity

$$(\text{Right Side})' = x \cdot (\ln x)^n + \frac{x^2}{2}\cdot n(\ln x)^{n-1}\cdot \frac{1}{x} - \frac{n}{2}x(\ln x)^{n-1}$$

$$= x \cdot (\ln x)^n \quad \text{(after a bit of algebra.)}$$

35. Let $u = (\ln x)^n$ and $dv = dx$.

37. (a) $-\dfrac{1}{2}\cos\left(x^2\right)+C$

    (b) $-\dfrac{1}{2}x^2\cos\left(x^2\right)+\dfrac{1}{2}\sin\left(x^2\right)+C$

    (c) $\dfrac{1}{2}x^2\sin\left(x^2\right)+\dfrac{1}{2}\cos\left(x^2\right)+C$

    (d) $\displaystyle\int x^2\cos\left(x^2\right)\,dx=\dfrac{1}{2}x\sin\left(x^2\right)-\dfrac{1}{2}\int\cos\left(x^2\right).$ Since the expression on the left side of the equals sign is not elementary, neither is the expression on the right side.

39. An integration by parts with $u=f(x)$ and $dv=\sin x\,dx$ shows that

$$\int_0^\pi f(x)\sin x\,dx=-f(x)\cos x\Big]_0^\pi+\int_0^\pi f'(x)\cos x\,dx=f(\pi)+f(0)+\int_0^\pi f'(x)\cos x\,dx.$$

An integration by parts with $u=\sin x$ and $dv=f''(x)\,dx$ shows that

$$\int_0^\pi f''(x)\sin x\,dx=f'(x)\sin x\Big]_0^\pi-\int_0^\pi f'(x)\cos x\,dx=-\int_0^\pi f'(x)\cos x\,dx.$$

Combining these results, we have

$$6=\int_0^\pi f(x)\sin x\,dx+\int_0^\pi f''(x)\sin x\,dx=f(\pi)+f(0)=f(\pi)+2.$$

From this it follows that $f(\pi)=4$.

9.2   Partial Fractions

1.   (b) $\displaystyle\int \frac{5x+7}{(x+1)(x+2)}\,dx = 2\ln|x+1| + 3\ln|x+2| + C$

3.   (a) Using partial fractions, $\dfrac{x^2+3x-1}{x(x+1)(x-2)} = \dfrac{1}{2}\cdot\dfrac{1}{x} - \dfrac{1}{x+1} + \dfrac{3}{2}\cdot\dfrac{1}{x-2}$. Thus, $A = 1/2$, $B = -1$, and $C = 3/2$.

    (b) $\displaystyle\int \frac{x^2+3x-1}{x(x+1)(x-2)}\,dx = \frac{1}{2}\ln|x| - \ln|x+1| + \frac{3}{2}\ln|x-2| + C.$

5. Let $I = \displaystyle\int \frac{6}{(x-2)(x^2-1)}\,dx$. Partial fractions work, again:

$$\frac{6}{(x-2)(x^2-1)} = \frac{6}{(x-2)(x-1)(x+1)} = \frac{A}{x-2} + \frac{B}{x-1} + \frac{C}{x+1}$$

Solving gives $A = 2$, $B = -3$, and $C = 1$, so $I = 2\ln|x-2| - 3\ln|x-1| + \ln|x+1| + K$.

7.   (a) $\dfrac{x^2+x}{(x^2+1)^2} = \dfrac{1}{x^2+1} + \dfrac{x-1}{(x^2+1)^2}$ (i.e., $A = 0$, $B = 1$, $C = 1$, $D = -1$)

    (b) $\displaystyle\int \frac{x^2+x}{(x^2+1)^2}\,dx = \frac{1}{2}\arctan x - \frac{x+1}{2(x^2+1)}$

9. $\displaystyle\int \frac{dx}{x^3+1} = \int \frac{dx}{(x+1)(x^2-x+1)} = \frac{1}{3}\int \left(\frac{1}{x+1} + \frac{2-x}{x^2-x+1}\right)dx =$

   $\dfrac{1}{3}\ln|x+1| - \dfrac{1}{6}\ln\left|x^2-x+1\right| + \dfrac{1}{2}\displaystyle\int \frac{dx}{x^2-x+1} =$

   $\dfrac{1}{3}\ln|x+1| - \dfrac{1}{6}\ln\left|x^2-x+1\right| + \dfrac{1}{\sqrt{3}}\arctan\left(\dfrac{2x-1}{\sqrt{3}}\right)$. Thus, $\displaystyle\int_0^2 \frac{dx}{x^3+1} = \frac{1}{6}\left(\ln 3 + \sqrt{3}\pi\right)$.

11. $\displaystyle\int \frac{x+1}{(x-1)(x+2)}\,dx = \int \left(\frac{2}{3}\cdot\frac{1}{x-1} + \frac{1}{3}\cdot\frac{1}{x+2}\right)dx = \frac{2}{3}\ln|x-1| + \frac{1}{3}\ln|(x+2| + C$

13. $\displaystyle\int \frac{5x^2+3x-2}{x^3+2x^2}\,dx = \int \left(-\frac{1}{x^2} + \frac{2}{x} + \frac{3}{x+2}\right)dx = \frac{1}{x} + 2\ln|x| + 3\ln|x+2| + C$

15. First complete the square: $I = \displaystyle\int \frac{x}{x^2+2x+6}\,dx = \int \frac{x}{(x+1)^2+5}\,dx$. Now let $u = x+1$, $x = u-1$, $dx = du$. Then,

$$I = \int \frac{u-1}{u^2+5}\,du = \int \frac{u}{u^2+5}\,du - \int \frac{du}{u^2+5}$$

$$= \frac{1}{2}\ln\left|u^2+5\right| - \frac{1}{\sqrt{5}}\arctan\left(\frac{u}{\sqrt{5}}\right) + C$$

$$= \frac{1}{2}\ln\left|x^2+2x+6\right| - \frac{1}{\sqrt{5}}\arctan\left(\frac{x+1}{\sqrt{5}}\right) + C$$

17. $\displaystyle\int \frac{x^3}{x^2+1}\,dx = \int \left(x - \frac{x}{x^2+1}\right)dx = \frac{1}{2}x^2 - \frac{1}{2}\ln(x^2+1) + C$

19. $\displaystyle\int \frac{3x^2-1}{(x-1)(x+2)}\,dx = \int \left(3 + \frac{2}{3}\cdot\frac{1}{x-1} - \frac{11}{3}\cdot\frac{1}{x+2}\right)dx = 3x + \frac{2}{3}\ln|x-1| - \frac{11}{3}\ln|x+2| + C$

21. $\displaystyle\int \frac{dx}{x\sqrt{x+1}} \to 2\int \frac{du}{u^2-1} = \int \left(\frac{1}{u-1} - \frac{1}{u+1}\right) du = \ln\left|\frac{u-1}{u+1}\right| + C \to \ln\left|\frac{\sqrt{x+1}-1}{\sqrt{x+1}+1}\right| + C$

23. $\displaystyle\int \frac{dx}{1+e^x} \to \int \frac{du}{u(u-1)} = \int \left(\frac{1}{u-1} - \frac{1}{u}\right) du = \ln\left|\frac{u-1}{u}\right| + C \to \ln\left(\frac{e^x}{1+e^x}\right) + C$

25. No. Since q is a quadratic polynomial, $\displaystyle\frac{q(x)}{(1-x)^2(3+x)} = \frac{P}{1-x} + \frac{Q}{(1-x)^2} + \frac{R}{3+x}$, where P, Q, and R are

   real numbers. Therefore, $\displaystyle\int \frac{q(x)}{(1-x)^2(3+x)}\, dx = -P\ln|1-x| + \frac{Q}{1-x} + R\ln(3+x) + S$, where P, Q, R,

   and S are real numbers. Since no choice of values for the constants P, Q, R, and S in the expression above

   leads to the expression $\displaystyle\frac{1}{1-x} + \arcsin x + \ln(3+x) + C$, we conclude that there is no quadratic polynomial q

   with the desired property.

27. (a) $\displaystyle(\ln|x+a|)' = \frac{1}{x+a}$

   (b) $\displaystyle\left(\frac{1}{1-n}\frac{1}{(x+a)^{n-1}}\right)' = \frac{1}{(x+a)^n}$

29. (a) Integration by parts with $u = 1/(x^2+a^2)^n$ and $dv = dx$ leads to the desired identity.

   (b) $\displaystyle\int \frac{x^2}{(x^2+a^2)^{n+1}}\, dx = \int \frac{(x^2+a^2) - a^2}{(x^2+a^2)^{n+1}}\, dx = \int \frac{dx}{(x^2+a^2)^n} - a^2 \int \frac{dx}{(x^2+a^2)^{n+1}}$

   (c) Rearranging the terms in part (b) leads to the equation

   $$2na^2 \int \frac{dx}{(x^2+a^2)^{n+1}} = \frac{x}{(x^2+a^2)^n} + (2n-1)\int \frac{dx}{(x^2+a^2)^n}.$$

   The desired identity follows from dividing both sides of the equation above by $2na^2$.

9.3 Trigonometric Antiderivatives

1. No—the two answers are equal. To see this, use the identity $\cos^2 x = 1 - \sin^2 x$.

3. $\displaystyle\int \cos^2(x/3)\,dx = \frac{1}{2}\int \big(1+\cos(2x/3)\big)\,dx = \frac{1}{2}x + \frac{3}{4}\sin(2x/3) + C$

5. $\displaystyle\int \cos^2 x \sin^3 x\,dx = \int \left(\cos^2 x - \cos^4 x\right)\sin x\,dx = -\frac{1}{3}\cos^3 x + \frac{1}{5}\cos^5 x + C$

7. $\displaystyle\int \sin^2 x \cos^2 x\,dx = \int \left(\sin^2 x - \sin^4 x\right) = \frac{1}{8}x + \frac{1}{8}\cos x \sin x - \frac{1}{4}\cos^3 x \sin x + C$

9. $\displaystyle\int \frac{\sin^3 x}{\cos x}\,dx = \int \frac{(1-\cos^2 x)\sin x}{\cos x}\,dx = \frac{1}{2}\cos^2 x - \ln|\cos x| + C$

11. $\displaystyle\int \frac{dx}{(x^2+4)^2} \;\rightarrow\; \frac{1}{8}\int \frac{\sec^2 t}{(1+\tan^2 t)^2}\,dt = \frac{1}{8}\int \cos^2 t\,dt = \frac{t}{16} + \frac{\sin(2t)}{32} + C$

$\displaystyle = \frac{t}{16} + \frac{\sin t \cos t}{16} + C \rightarrow \frac{1}{16}\arctan(x/2) + \frac{x}{8\,(4+x^2)} + C$

$[x = 2\tan t,\, dx = 2\sec^2 t\,dt,\, \sin t = x/\sqrt{4+x^2},\, \cos t = 2/\sqrt{4+x^2}]$

13. $\displaystyle\int \frac{x^2}{\sqrt{9-x^2}}\,dx \;\rightarrow\; 9\int \sin^2 t\,dt = \frac{9t}{2} - \frac{9}{4}\sin(2t) + C = \frac{9t}{2} - \frac{9}{2}\sin t \cos t + C$

$\displaystyle\rightarrow\; \frac{9}{2}\arcsin(x/3) - \frac{1}{2}x\sqrt{9-x^2} + C$

$[x = 3\sin t,\, dx = 3\cos t\,dt,\, \cos t = \sqrt{1-x^2/9}]$

15. $\displaystyle\int \frac{dx}{\sqrt{1+x^2}} \rightarrow \int \frac{\sec^2 t}{\sqrt{1+\tan^2 t}}\,dt = \int \sec t\,dt = \ln|\tan t + \sec t| + C \rightarrow \ln\left|x+\sqrt{1+x^2}\right| + C$

17. $\displaystyle\int \tan^4 x\,dx = \frac{1}{3}\tan^3 x - \tan x + x + C$

19. $\displaystyle\int \sec^3 x \tan^2 x\,dx = \int \left(\sec^5 x - \sec^3 x\right)\,dx = \frac{1}{4}\sec^3 x \tan x - \frac{1}{8}\sec x \tan x - \frac{1}{8}\ln(\sec x + \tan x) + C$

21. $\displaystyle\int \sqrt{\cos x}\,\sin^5 x\,dx = \int \sqrt{\cos x}\left(1-\cos^2 x\right)^2 \sin x\,dx = -\frac{2}{3}(\cos x)^{3/2} + \frac{4}{7}(\cos x)^{7/2} - \frac{2}{11}(\cos x)^{11/2} + C$

23. $\displaystyle\int \sqrt{1+x^2}\,dx \rightarrow \int \sec t \sec^2 t\,dt = \int \sec^3 t\,dt = \frac{1}{2}\tan t \sec t + \frac{1}{2}\ln|\sec t + \tan t| + C$

$\displaystyle\rightarrow\; \frac{1}{2}x\sqrt{1+x^2} + \frac{1}{2}\ln\left|\sqrt{1+x^2}+x\right| + C$

$[x = \tan t,\, dx = \sec^2 t,\, \sec t = \sqrt{1+x^2}]$

25. $\displaystyle\int \frac{dx}{x^2\sqrt{x^2-4}} \rightarrow \frac{1}{4}\int \cos t\,dt = \frac{1}{4}\sin t + C \rightarrow \frac{\sqrt{x^2-4}}{4x} + C$

$[x = 2\sec t,\, dx = 2\sec t \tan t\,dt,\, \sin t = \sqrt{1-4/x^2}]$

27. First, note that $\displaystyle\int \frac{x+2}{x(x^2+1)}\,dx = \int \frac{dx}{x^2+1} + 2\int \frac{dx}{x(x^2+1)} = \arctan x + 2\int \frac{dx}{x(x^2+1)}$. Also,

$\displaystyle\int \frac{dx}{x(x^2+1)} \rightarrow \int \frac{\cos t}{\sin t}\,dt = \ln|\sin t| + C \rightarrow \ln\left|\frac{x}{\sqrt{1+x^2}}\right| + C.$

$[x = \tan t, dx = \sec^2 t, \sin t = x/\sqrt{1 + x^2}]$

Therefore, $\displaystyle\int \frac{x + 2}{x(x^2 + 1)} \, dx = \arctan x + 2 \ln \left| \frac{x}{\sqrt{1 + x^2}} \right| + C.$

29. (a) $\cos(x + y) + \cos(x - y) = (\cos x \cos y - \sin x \sin y) + (\cos x \cos(-y) - \sin x \sin(-y)) = 2 \cos x \cos y$

(b) $\displaystyle\int \cos(ax) \cos(bx) \, dx = \frac{1}{2} \int \left(\cos((a + b)x) + \cos((a - b)x) \right) dx = \frac{1}{2(a + b)} \sin((a + b)x) +$

$\displaystyle\frac{1}{2(a - b)} \sin((a - b)x) + C$

(c) $\displaystyle\int \cos(ax) \cos(ax) \, dx = \int \cos^2(ax) \, dx = \frac{x}{2} + \frac{1}{2a} \cos(ax) \sin(ax) + C$

31. (a) $\sin(x + y) + \sin(x - y) = (\sin x \cos y + \cos x \sin y) + (\sin x \cos(-y) + \cos x \sin(-y)) = 2 \sin x \cos y.$

(b) $\displaystyle\int \sin(ax) \cos(bx) \, dx = \frac{1}{2} \int \left(\sin((a + b)x) + \sin((a - b)x) \right) dx = -\frac{1}{2(a + b)} \cos((a + b)x) - $

$\displaystyle\frac{1}{2(a - b)} \cos((a - b)x) + C$

(c) $\displaystyle\int \sin(ax) \cos(ax) \, dx = \frac{1}{2a} \sin^2(ax) + C$

33. (a) $u = \sin^{n-1} x \implies du = (n - 1) \sin^{n-2} x \cos x \, dx; dv = \sin x \, dx \implies v = -\cos x.$ Thus,

$$\int \sin^n x \, dx = \int u \, dv = uv - \int v \, du = -\sin^{n-1} x \cos x + (n - 1) \int \sin^{n-2} x \cos^2 x \, dx.$$

(b) $\displaystyle\int \sin^{n-2} x \cos^2 x \, dx = \int \sin^{n-2} x \left(1 - \sin^2 x \right) dx = \int \sin^{n-2} x \, dx - \int \sin^n x \, dx.$ Thus,

$$\int \sin^n x \, dx = -\sin^{n-1} x \cos x + (n - 1) \int \sin^{n-2} x \, dx - (n - 1) \int \sin^n x \, dx$$

so $\displaystyle n \int \sin^n x \, dx = -\sin^{n-1} x \cos x + (n - 1) \int \sin^{n-2} x \, dx.$ Therefore, if $n \neq 0,$

$$\int \sin^n x \, dx = -\frac{\sin^{n-1} x \cos x}{n} + \frac{n - 1}{n} \int \sin^{n-2} x \, dx.$$

35. The absolute value is required because $\sqrt{x^2 - 4} > 0$ when $x < -2,$ but $2 \tan t < 0$ when $\pi/2 < t < \pi.$

37. $\displaystyle\int \frac{dx}{1 + \cos x} \to \int \frac{1}{1 + (1 - t^2)/(1 + t^2)} \frac{2}{1 + t^2} \, dt = \int dt = t + C \to \tan(x/2) + C$

9.4 Miscellaneous Exercises

1. $\int \dfrac{\sin x}{(3 + \cos x)^2}\, dx = \dfrac{1}{3 + \cos x}$

[substitution: $u = 3 + \cos x$]

3. $\int x\left(3 + 4x^2\right)^5 dx = \dfrac{1}{48}\left(3 + 4x^2\right)^6$

[subsitution: $u = 3 + 4x^2$]

5. $\int \dfrac{x}{\sqrt[3]{x^2 + 4}}\, dx = \dfrac{3}{4}\left(x^2 + 4\right)^{2/3}$

[subsitution: $u = x^2 + 4$]

7. $\int x e^x\, dx = e^x(x - 1)$

[integration by parts: $u = x,\, dv = e^x\, dx$]

9. $\int \dfrac{\ln x}{x}\, dx = \dfrac{1}{2}\left(\ln |x|\right)^2$

[substituion: $u = \ln x$]

11. $\int \sin^2(3x)\cos(3x)\, dx = \dfrac{1}{9}\sin^3(3x)$

[subsitution: $u = \sin(3x)$]

13. $\int x e^{3x^2}\, dx = \dfrac{1}{6}e^{3x^2}$

[substitution: $u = 3x^2$]

15. $\int \dfrac{7 - x}{(x + 3)(x^2 + 1)}\, dx = \ln |x + 3| + 2\arctan x - \dfrac{1}{2}\ln(x^2 + 1)$

[partial fractions: $\frac{7-x}{(x+3)(x^2+1)} = \frac{1}{x+3} + \frac{2-x}{x^2+1}$]

17. $\int \arctan x\, dx = x\arctan x - \dfrac{1}{2}\ln(1 + x^2)$

[integration by parts: $u = \arctan x,\, dv = dx$]

19. $\int x\sin x\, dx = \sin x - x\cos x$

[integration by parts: $u = x,\, dv = \sin x\, dx$]

21. $\int x^2 \ln x\, dx = \dfrac{1}{3}x^3 \ln |x| - \dfrac{1}{9}x^3$

[integration by parts: $u = \ln x,\, dv = x^2\, dx$]

23. $\int \dfrac{e^x}{\sqrt{1 - e^{2x}}}\, dx = \arcsin\left(e^x\right)$

[substitution: $u = e^x$]

25. $\int \ln x\, dx = x(\ln x - 1)$

[integration by parts: $u = \ln x,\, dv = dx$]

27. $\int \arcsin x\, dx = x\arcsin x + \sqrt{1 - x^2}$

[integration by parts: $u = \arcsin x,\, dv = dx$]

29. $\int \frac{x}{\sqrt{x-2}}\, dx = \frac{2}{3}(x-2)^{3/2} + 4\sqrt{x-2}$

[substitution: $u = x - 2$]

31. $\int \frac{x+6}{(x+1)\left(x^2+4\right)}\, dx = \ln|x+1| + \arctan(x/2) - \frac{1}{2}\ln(x^2+4)$

[partial fractions: $\frac{x+6}{(x+1)(x^2+4)} = \frac{1}{x+1} + \frac{2-x}{x^2+4}$]

33. $\int \frac{x^3}{1+x^2}\, dx = \frac{1}{2}(1+x^2) - \frac{1}{2}\ln(1+x^2)$

[substitution: $u = 1 + x^2$]

35. $\int \cos(2x)\, dx = \frac{1}{2}\sin(2x)$

37. $\int \frac{dx}{1+x^2} = \arctan x$

39. Let $u = \sin x$. Then, $du = \cos x\, dx$ and

$$\int \sin^3 x \cos^3 x\, dx = \int \sin^3 x \cos^2 x \cos x\, dx = \int \sin^3 x(1 - \sin^2 x)\cos x\, dx$$

$$= \int \sin^3 x \cos x\, dx - \int \sin^5 x \cos x\, dx \rightarrow \int u^3\, du - \int u^5\, du$$

$$= \tfrac{1}{4}u^4 - \tfrac{1}{6}u^6 + C \rightarrow \tfrac{1}{4}\sin^4 x - \tfrac{1}{6}\sin^6 x + C$$

41. Let $u = \ln x$. Then, $du = dx/x$ and $\int \frac{dx}{x(\ln x)^2} \rightarrow \int \frac{du}{u^2} = -\frac{1}{u} + C \rightarrow -\frac{1}{\ln|x|} + C.$

43. $\int \frac{dx}{x^3 + x} = \ln|x| - \frac{1}{2}\ln(x^2 + 1)$

[partial fractions: $\frac{1}{x^3+x} = \frac{1}{x} - \frac{x}{x^2+1}$]

45. $\int \frac{x+5}{x^2+3x-4}\, dx = \frac{6}{5}\ln|x-1| - \frac{1}{5}\ln|x+4|$

[partial fractions: $\frac{x+5}{x^2+3x-4} = \frac{6}{5(x-1)} - \frac{1}{5(x-4)}$]

47. $\int \frac{dx}{\sqrt[3]{x-1}} = \frac{3}{2}(x-1)^{2/3}$

49. $\int x^3 e^{x^2}\, dx = \frac{1}{2}e^{x^2}(x^2 - 1)$

[substitution ($w = x^2$), then integration by parts ($u = w,\ dv = e^w\, dw$)]

51. $\int \frac{dx}{2x - x^2} = \frac{1}{2}\ln|x| - \frac{1}{2}\ln|x-2| = \frac{1}{2}\ln\left|\frac{x}{x-2}\right|$

[partial fractions: $\frac{1}{2x-x^2} = \frac{1}{x(2-x)} = \frac{1}{2x} + \frac{1}{2(2-x)}$]

53. $\int \left(x^2 + 2x + 3\right)^{3/2} dx = \frac{1}{4}(x+1)(x^2+2x+3)^{3/2} + \frac{3}{4}(x+1)\sqrt{x^2+2x+3} +$
$\frac{3}{2}\ln\left|\frac{\sqrt{x^2+2x+3}}{\sqrt{2}} + \frac{x+1}{\sqrt{2}}\right|$

[Write $x^2 + 2x + 3 = (x+1)^2 + 2$, then use a trigonometric substitution ($x + 1 = \sqrt{2}\tan t$).]

55. $\displaystyle \int e^x e^{2x}\, dx = \int e^{3x}\, dx = \frac{1}{3}e^{3x}$

57. $\displaystyle \int \ln(1+x^2)\, dx = x\ln(1+x^2) - 2x + 2\arctan x$

[integration by parts: $u = \ln(1+x^2),\ dv = dx$]

59. $\displaystyle \int \frac{x}{9+4x^4}\, dx = \frac{1}{12}\arctan\left(\frac{2x^2}{3}\right)$

61. $\displaystyle \int \sin(3x)\cos(5x)\, dx = \frac{1}{4}\cos(2x) - \frac{1}{16}\cos(8x)$

[Write $\sin(3x)\cos(5x) = \frac{1}{2}\sin(8x) - \frac{1}{2}\sin(2x)$.]

63. $\displaystyle \int x\sqrt{2x+1}\, dx = \frac{1}{10}(2x+1)^{5/2} - \frac{1}{6}(2x+1)^{3/2}$

[substitution: $u = 2x+1$]

65. $\displaystyle \int \frac{\tan x}{\sec^2 x}\, dx = -\frac{1}{2}\cos^2 x$

[Write $\frac{\tan x}{\sec^2 x} = \sin x \cos x$, then use substitution ($u = \cos x$).]

67. $\displaystyle \int \frac{x}{16+9x^2}\, dx = \frac{1}{18}\ln\left(16+9x^2\right)$

[subsitution: $u = 16+9x^2$]

69. $\displaystyle \int \frac{dx}{(e^x - e^{-x})^2} = \frac{1}{4}\left(\frac{1}{1+e^x} + \frac{1}{1-e^x} = \frac{1}{2-2e^{2x}}\right)$

[Write $\left(e^x - e^{-x}\right)^{-2} = e^{2x}\left(e^{2x}-1\right)^{-2} = (e^x)^2\,(e^x-1)^{-2}\,(e^x+1)^{-2}$, then use substitution ($u = e^x$) and partial fractions.]

71. $\displaystyle \int x\tan^2 x\, dx = x\tan x - \frac{1}{2}x^2 - \ln|\sec x|$

[integration by parts: $u = x,\ dv = \tan^2 x\, dx$]

73. $\displaystyle \int \frac{x^3}{(x^2+1)^2} = \frac{1}{2(x^2+1)} + \frac{1}{2}\ln(x^2+1)$

[substitution $u = x^2+1$]

75. $\displaystyle \int \sin x \sin(2x)\, dx = \frac{1}{2}\sin x - \frac{1}{6}\sin(3x)$

[Write $\sin x \sin(2x) = \frac{1}{2}\cos x - \frac{1}{2}\cos(3x)$.]

77. $\displaystyle \int \frac{x}{1+x^4}\, dx = \frac{1}{2}\arctan(x^2)$

[substitution: $u = x^2$]

79. $\displaystyle \int \sin^5 x \cos^2 x\, dx = \frac{1}{7}\sin^6 x \cos x - \frac{1}{35}\sin^4 x \cos x - \frac{4}{105}\sin^2 x \cos x - \frac{8}{105}\cos x$

[Write $\cos^2 x = 1 - \sin^2 x$, then use a reduction formula.]

10.1   When Is an Integral Improper?

1. The interval of integration is infinite.

3. The integrand is unbounded as $x \to 1^+$.

5. The integrand is unbounded as $x \to \pi/2^-$

7. $\displaystyle\int_0^\infty \frac{dx}{x^2} = \int_0^1 \frac{dx}{x^2} + \int_1^\infty \frac{dx}{x^2}$. Since $\displaystyle\int_0^1 \frac{dx}{x^2} = \infty$, the original improper integral diverges.

9. $\displaystyle\int_0^\infty e^{-x}\,dx = \lim_{t\to\infty} \int_0^t = \lim_{t\to\infty} \left. -e^{-x} \right]_0^t = \lim_{t\to\infty} \left(1 - e^{-t}\right) = 1$

11. $\displaystyle\int_1^\infty \frac{dx}{x(1+x)} = \lim_{t\to\infty} \left. \ln\left(\frac{x}{1+x}\right)\right]_1^t = \lim_{t\to\infty} \ln\left(\frac{t}{1+t}\right) - \ln\tfrac{1}{2} = \ln 2$

13. $\displaystyle\int_{-2}^2 \frac{2x+1}{\sqrt[3]{x^2+x-6}}\,dx = \lim_{t\to2^-} \int_{-2}^t \frac{2x+1}{\sqrt[3]{x^2+x-6}}\,dx = \lim_{t\to2^-} \frac{3}{2}\left. \left(x^2+x-6\right)^{2/3}\right]_{-2}^t$

$\displaystyle \qquad = \lim_{t\to2^-} \frac{3}{2}\left(\left(t^2+t-6\right)^{2/3} - (-4)^{2/3}\right) = \tfrac{3}{2}\sqrt[3]{16}$

15. $\displaystyle\int_0^\infty \frac{\arctan x}{(1+x^2)^{3/2}} \to \int_0^{\pi/2} w\cos w\,dw = \left. w\sin w + \cos w \right]_0^{\pi/2} = \frac{\pi}{2} - 1.$

NOTE: $\displaystyle\int \frac{\arctan x}{(1+x^2)^{3/2}} \to \int w\cos w\,dw = w\sin w + \cos w \to \frac{1 + x\arctan x}{\sqrt{1+x^2}}.$

17. $\displaystyle\int_a^\infty e^{-x}\,dx = e^{-a} \le 10^{-5}$ if $a \ge 11.6$.

19. $\displaystyle\int_a^\infty \frac{dx}{x(\ln x)^3} = \frac{1}{2(\ln a)^2} \le 10^{-5}$ if $a \ge 5 \times 10^{97}$.

21. $\displaystyle I = \int_0^\infty f(x)\,dx = \int_0^a f(x)\,dx + \int_a^\infty f(x)\,dx \implies \left| I - \int_0^a f(x)\,dx \right| = \left| \int_a^\infty f(x)\,dx \right| \le 0.0001.$

23. (a) The integral is improper because the integrand is unbounded near $x = 0$.

   (b) No — $\displaystyle\int_{-1}^1 x^{-3}\,dx = \int_{-1}^0 x^{-3}\,dx + \int_0^1 x^{-3}\,dx$ and both of the latter improper integrals diverge.

25. Converges. $\displaystyle\int_0^\infty \frac{\arctan x}{1+x^2}\,dx = \lim_{t\to\infty} \frac{1}{2}(\arctan t)^2 = \pi^2/8$

27. Converges. $\displaystyle\int_3^\infty \frac{x}{(x^2-4)^3}\,dx = \lim_{t\to\infty} \frac{1}{4}\left(\frac{1}{25} - \frac{1}{t^2-4}\right) = \frac{1}{100}$

29. Converges. $\displaystyle\int_0^8 \frac{dx}{\sqrt[3]{x}} = \lim_{t\to0^+} \frac{3}{2}\left(8^{2/3} - t^{2/3}\right) = 6$

31. Converges. $\displaystyle\int_2^3 \frac{x}{\sqrt{3-x}}\,dx = \lim_{t\to3^-}\left(\frac{2}{3}(3-t)^{3/2} - 6\sqrt{3-t} + \frac{16}{3}\right) = \frac{16}{3}$

33. Diverges. $\displaystyle\int_1^\infty \frac{dx}{x(\ln x)^2} = \int_1^2 \frac{dx}{x(\ln x)^2} + \int_2^\infty \frac{dx}{x(\ln x)^2} = \lim_{s\to1}\left(\frac{1}{\ln s} - \frac{1}{\ln 2}\right) + \lim_{t\to\infty}\left(\frac{1}{\ln 2} - \frac{1}{\ln t}\right) = \infty$

35. Diverges. $\displaystyle\int_0^\infty \frac{dx}{e^x - 1} = \int_0^1 \frac{dx}{e^x - 1} + \int_1^\infty \frac{dx}{e^x - 1}$

37. Converges. $\displaystyle\int_{-\infty}^\infty \frac{dx}{e^x + e^{-x}} = \int_{-\infty}^\infty \frac{e^x}{e^{2x} + 1}\, dx = \frac{\pi}{2}$

39. Converges. $\displaystyle\int_0^1 \frac{e^{-\sqrt{x}}}{\sqrt{x}}\, dx = 2 - 2/e$

41. (a) $\displaystyle\int_1^\infty \frac{dx}{x} = \lim_{t\to\infty} \int_1^t \frac{dx}{x} = \lim_{t\to\infty} \ln t = \infty$

  (b) If $p > 1$, $\displaystyle\int_1^\infty \frac{dx}{x^p} = \lim_{t\to\infty} \frac{1 - t^{1-p}}{p - 1} = \frac{1}{p-1}$.

  (c) If $p < 1$, $\displaystyle\int_1^\infty \frac{dx}{x^p} = \lim_{t\to\infty} \frac{1 - t^{1-p}}{p - 1} = \infty$.

43. (a) $p < 1$

  (b) $p > 1$

45. $\displaystyle\int \left(\frac{2x}{x^2 + 1} - \frac{C}{2x + 1} \right) dx = \ln(x^2 + 1) - \frac{C}{2} \ln(2x + 1) = \ln\left(\frac{x^2 + 1}{\sqrt{(2x + 1)^C}} \right)$. Thus, the improper integral
converges to $-2 \ln 2$ if $C = 4$ and diverges for all other values of C.

47. $\displaystyle\int \left(\frac{Cx^2}{x^3 + 1} - \frac{1}{3x + 1} \right) dx = \frac{C}{3} \ln(x^3 + 1) - \frac{1}{3} \ln(3x + 1)$. Thus, the improper integral converges to $-\frac{1}{3} \ln 3$
if $C = \frac{1}{3}$ and diverges for all other values of C.

49. $\displaystyle\int_0^\infty \left(\frac{x}{x^2 + 1} - \frac{C}{3x + 1} \right) dx = \frac{1}{2} \ln\left(x^2 + 1 \right) - \frac{C}{3} \ln(3x + 1)$. Thus, the improper integral converges to $-\ln 3$
if $C = 3$ and diverges for all other values of C.

51. $\displaystyle\int_1^\infty \frac{x}{x^3 + 1}\, dx = \lim_{t\to\infty} \int_1^t \frac{x}{x^3 + 1}\, dx = \lim_{t\to\infty} - \int_1^{1/t} \frac{du}{1 + u^3} = \int_0^1 \frac{du}{1 + u^3}$

53. Use the substition $u = 1/x$:

$$\int_0^\infty \frac{dx}{1 + x^4} = \int_0^1 \frac{dx}{1 + x^4} + \int_1^\infty \frac{dx}{1 + x^4} = \lim_{s\to 0^+} \int_s^1 \frac{dx}{1 + x^4} + \lim_{t\to\infty} \int_1^t \frac{dx}{1 + x^4}$$

$$= \lim_{s\to 0^+} - \int_{1/s}^1 \frac{u^2}{u^4 + 1}\, du + \lim_{t\to\infty} - \int_1^{1/t} \frac{u^2}{u^4 + 1}\, du$$

$$= \int_1^\infty \frac{u^2}{u^4 + 1}\, du + \int_0^1 \frac{u^2}{u^4 + 1}\, du = \int_0^\infty \frac{u^2}{u^4 + 1}\, du$$

55. $\displaystyle\int_0^\infty \frac{x \ln x}{1 + x^4}\, dx = 0$.   Using the substitution $u = 1/x$: $\displaystyle\int_1^\infty \frac{x \ln x}{1 + x^4}\, dx = -\int_0^1 \frac{u \ln u}{u^4 + 1}\, dx$.

10.2 Detecting Convergence, Estimating Limits

1. (a) For every $x \in \mathbb{R}$, $-1 \leq \sin x \leq 1$

   (b) $\int_2^\infty \dfrac{dx}{x + \sin x} \geq \int_2^\infty \dfrac{dx}{x + 1} = \infty$

3. (a) $0 \leq \sqrt{x} \leq x^2/2$ for all $x \geq 2$, so $x^2/2 = x^2 - x^2/2 \leq x^2 - \sqrt{x} \leq x^2$ for all $x \geq 2$.

   (b) $0 \leq \int_3^\infty \dfrac{dx}{x^2 - \sqrt{x}} \leq 2 \int_3^\infty \dfrac{dx}{x^2} = \dfrac{1}{3}$

5. (a) When $x \geq 1$, $\sqrt{x^3} \leq \sqrt{1 + x^3} \leq \sqrt{2x^3}$. Therefore,

$$\frac{1}{\sqrt{2x}} = \frac{x}{\sqrt{2x^3}} \leq \frac{x}{\sqrt{1 + x^3}} \leq \frac{x}{\sqrt{x^3}} = \frac{1}{\sqrt{x}}$$

   when $x \geq 1$.

   (b) The improper integral **diverges** since

$$
\begin{aligned}
0 \leq \int_0^\infty \frac{x}{\sqrt{1 + x^3}}\, dx &= \int_0^1 \frac{x}{\sqrt{1 + x^3}}\, dx + \int_1^\infty \frac{x}{\sqrt{1 + x^3}}\, dx \\
&\geq \int_0^1 \frac{x}{\sqrt{1 + x^3}}\, dx + \int_1^\infty \frac{dx}{\sqrt{2x}}
\end{aligned}
$$

7. (a) When $x \geq 1$, $\sqrt{x} \geq 1$. Therefore, $\sqrt{x} \cdot x^4 = \sqrt{x} \cdot \sqrt{x^8} = \sqrt{x^9} < \sqrt{1 + x^9}$.
   When $x \geq 1$, $x^9 \geq 1$. Therefore, $\sqrt{1 + x^9} \leq \sqrt{x^9 + x^9} = \sqrt{2}\, x^{9/2}$.

   (b) Converges. $0 \leq \int_0^\infty \dfrac{dx}{\sqrt{1 + x^9}} \leq \int_0^1 \dfrac{dx}{\sqrt{1 + x^9}} + \int_1^\infty \dfrac{dx}{x^{9/2}} < \infty$

9. $0 < \int_a^\infty \dfrac{dx}{x^4\sqrt{2x^3 + 1}} < \dfrac{1}{\sqrt{2}} \int_a^\infty \dfrac{dx}{x^{11/2}} = \dfrac{\sqrt{2}}{9} a^{-9/2} \leq 10^{-5}$ if $a \geq \left(\dfrac{2 \times 10^{10}}{81}\right)^{1/9} \approx 8.5606$. Thus,

   $\int_1^9 \dfrac{dx}{x^4\sqrt{2x^3 + 1}}$ approximates $\int_1^\infty \dfrac{dx}{x^4\sqrt{2x^3 + 1}}$ within 10^{-5}.

11. $0 < \int_a^\infty \dfrac{e^{-x}}{2 + \cos x}\, dx < \int_a^\infty e^{-x}\, dx = e^{-a} \leq 10^{-5}$ if $a \geq \ln(100000) \approx 11.513$. Thus, $\int_0^{12} \dfrac{e^{-x}}{2 + \cos x}\, dx$

   approximates $\int_0^\infty \dfrac{e^{-x}}{2 + \cos x}\, dx$ within 10^{-5}.

13. (a) Let $f(x) = \sin x - x/2$. Then $f(0) = 0$ and $f'(x) = \cos x - 1/2 > 0$ if $0 \leq x \leq 1$. Therefore, $f(x) > 0$
    if $0 \leq x \leq 1$.

    (b) $0 \leq \int_0^1 \dfrac{dx}{\sqrt{\sin x}} \leq \sqrt{2} \int_0^1 \dfrac{dx}{\sqrt{x}} = 2\sqrt{2}$.

15. Converges. $0 \leq \int_1^\infty \dfrac{dx}{x^4 + 1}\, dx \leq \int_1^\infty \dfrac{dx}{x^4} = \dfrac{1}{3}$

17. Diverges. $\int_2^\infty \dfrac{dx}{\sqrt{x} - 1}\, dx > \int_2^\infty \dfrac{dx}{\sqrt{x}} = \infty$.

19. Converges. $0 \leq \int_1^\infty \dfrac{dx}{x\sqrt{1 + x}} \leq \int_1^\infty \dfrac{dx}{x^{3/2}} = 2$.

21. Diverges. $\displaystyle\int_0^\infty \frac{dx}{x+e^{-x}} = \int_0^1 \frac{dx}{x+e^{-x}} + \int_1^\infty \frac{dx}{x+e^{-x}} > \int_0^1 \frac{dx}{x+e^{-x}} + \int_1^\infty \frac{dx}{2x}$. Since the improper integral on the right diverges, the original improper integral diverges.

23. Diverges. $\displaystyle\int_1^\infty \frac{\sqrt{x}}{x+1}\,dx \geq \frac{1}{2}\int_1^\infty \frac{dx}{\sqrt{x}} = \infty$

25. Diverges. $\displaystyle\int_1^\infty \frac{dx}{1+\sqrt{x}} \geq \int_1^\infty \frac{dx}{2\sqrt{x}} = \infty$

27. Converges. $\displaystyle 0 \leq \int_0^\infty \frac{dx}{\sqrt{x}(1+x)} = \int_0^1 \frac{dx}{\sqrt{x}(1+x)} + \int_1^\infty \frac{dx}{\sqrt{x}(1+x)} \leq \int_0^1 \frac{dx}{\sqrt{x}} + \int_1^\infty \frac{dx}{x^{3/2}} = 2+2 = 4$

29. Let $u = 1/x$. Then, $x = u^{-1}$ and $-u^{-2}\,du = dx$ so

$$\int_0^1 \frac{dx}{\sqrt{x}(1+x)} = \lim_{t\to 0^+}\int_t^1 \frac{dx}{\sqrt{x}(1+x)} \to \lim_{t\to 0^+} -\int_{1/t}^1 \frac{du}{u^2\sqrt{u^{-1}+u^{-3}}}$$

$$= \lim_{t\to 0^+}\int_1^{1/t} \frac{du}{\sqrt{u^3+u}}$$

$$= \int_1^\infty \frac{du}{\sqrt{u^3+u}} = \int_1^\infty \frac{dx}{\sqrt{x+x^3}}$$

31. (a) f is an increasing function for all $x \geq 0$ since $f'(x) = \sqrt{x}e^{-x} > 0$ for these values of x. Thus, the limit exists if and only if f is bounded above. Since $f(x) = \int_3^x \sqrt{t}e^{-t}\,dt < \int_3^x te^{-t}\,dt = 4e^{-3} - (x+1)e^{-x}$, $\displaystyle\lim_{x\to\infty} f(x) < 4e^{-3}$. Therefore, the limit exists.

    (b) $\displaystyle\int_a^\infty \sqrt{t}e^{-t}\,dt \leq \int_a^\infty te^{-t}\,dt = (a+1)e^{-a}$ for all $a \geq 1$. Since $(a+1)e^{-a} < 0.001$ for all $a \geq 9.2335$, $f(a)$ approximates $\displaystyle\int_3^\infty \sqrt{x}e^{-x}\,dx$ within 0.001 when $a \geq 9.2335$.

33. $\displaystyle\int_a^\infty \frac{dx}{x^4+\sqrt{x}} \leq \int_a^\infty \frac{dx}{x^4} = \frac{1}{3}a^{-3} \leq 0.0025$ if $a \geq \sqrt[3]{400/3} \approx 5.1087$. Thus, $\displaystyle\int_1^6 \frac{dx}{x^4+\sqrt{x}}$ approximates $\displaystyle\int_1^\infty \frac{dx}{x^4+\sqrt{x}}$ with an error no greater than 0.0025.
Using $K_2 = 2.5$ and $K_4 = 21$, we find that a midpoint rule estimate computed with $n \geq 73$ (or a Simpson's rule estimate computed with $n \geq 20$) approximates $\displaystyle\int_1^6 \frac{dx}{x^4+\sqrt{x}}$ with an error no greater than 0.0025. Therefore, $M_{73} \approx 0.23603$ and $S_{20} \approx 0.23629$ are estimates of $\displaystyle\int_a^\infty \frac{dx}{x^4+\sqrt{x}}$ guaranteed to be accurate within 0.005.

35. $\displaystyle\int_a^\infty \frac{dx}{e^{x^2}+x} \leq \int_a^\infty e^{-x^2}\,dx \leq \frac{1}{2}e^{-a^2} \leq 0.0025$ if $a \geq \sqrt{\ln 200} \approx 2.3018$. Thus, $\displaystyle\int_0^3 \frac{dx}{e^{x^2}+x}$ approximates $\displaystyle\int_0^\infty \frac{dx}{e^{x^2}+x}$ with an error no greater than 0.0025.
Using $K_2 = 0.7$ and $K_4 = 36$, we find that the midpoint rule with $n \geq 18$ and Simpson's rule with $n \geq 12$ approximate $\displaystyle\int_0^3 \frac{dx}{e^{x^2}+x}$ with an error no greater than 0.0025. Therefore, $M_{18} \approx 0.69764$ and $S_{12} \approx 0.69870$ are estimates of $\displaystyle\int_0^\infty \frac{dx}{e^{x^2}+x}$ guaranteed to be accurate within 0.005.

37. (a) Since $\left| \dfrac{\cos x}{x^2} \right| \le \dfrac{1}{x^2}$ and $\displaystyle\int_1^\infty x^{-2}\, dx$ converges, Theorem 2 asserts that $\displaystyle\int_1^\infty \dfrac{\cos x}{x^2}\, dx$ converges.

(b) Let $u = x^{-1}$ and $dv = \sin x\, dx$. Then, $\displaystyle\int_1^\infty \dfrac{\sin x}{x}\, dx = -\dfrac{\cos x}{x}\Bigg]_1^\infty + \int_1^\infty \dfrac{\cos x}{x^2}\, dx = \cos 1 +$

$\displaystyle\int_1^\infty \dfrac{\cos x}{x^2}\, dx$. Thus, $\displaystyle\int_1^\infty \dfrac{\sin x}{x}\, dx$ can be written as the sum of two numbers (i.e., it converges).

(c) Let $w = e^x$. Then, $\displaystyle\int_0^\infty \sin\left(e^x\right)\, dx = \int_1^\infty \dfrac{\sin w}{w}\, dw$.

10.3 Improper Integrals and Probability

1. Graphs appear below; for reference, the the standard normal graph appears, too:

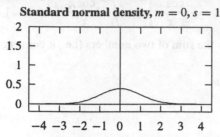

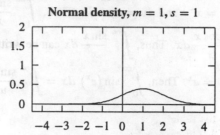

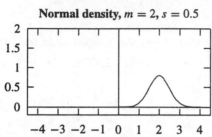

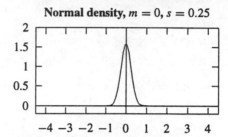

Notice the similarities among all the graphs—m locates the center; s determines the "spread."

3. When $Z = \dfrac{x - 500}{100}$, $dZ = \dfrac{dx}{100}$. Furthermore, when $x = 500$, $Z = 0$; when $x = 700$, $Z = 2$. Thus,

$$I_1 = \frac{1}{100\sqrt{2\pi}} \int_{500}^{700} \exp\left(-\frac{(x - 500)^2}{2 \cdot 100^2}\right) dx = \frac{1}{100\sqrt{2\pi}} \int_0^2 \exp\left(-\frac{Z^2}{2}\right) 100\, dZ = \frac{1}{\sqrt{2\pi}} \int_0^2 e^{-Z^2/2}\, dZ$$

5. (a) Since the integrand is an even function $\displaystyle\int_{-\infty}^{\infty} e^{-x^2}\, dx = 2 \int_0^{\infty} e^{-x^2}\, dx = 2 \cdot \dfrac{\sqrt{\pi}}{2} = \sqrt{\pi}.$

   (b) Using the substitution $u = x/\sqrt{2}$, $\dfrac{1}{\sqrt{2\pi}} \displaystyle\int_{-\infty}^{\infty} \exp\left(-x^2/2\right) = \dfrac{1}{\sqrt{\pi}} \int_{-\infty}^{\infty} e^{-u^2}\, du = 1.$

   (c) Using the substitution $u = (x - m)/s$, $\dfrac{1}{\sqrt{2\pi}\, s} \displaystyle\int_{-\infty}^{\infty} \exp\left(-(x - m)^2/2s^2\right) = \dfrac{1}{2\sqrt{\pi}} \int_{-\infty}^{\infty} e^{-u^2/2}\, du = 1.$

7. (a) $z = \dfrac{600 - 500}{100} = 1$

   (b) $z = \dfrac{450 - 500}{100} = -0.5$

   (c) $\dfrac{1}{\sqrt{2\pi} \cdot 100} \displaystyle\int_{450}^{\infty} \exp\left(-\frac{x - 500}{2 \cdot 100^2}\right) dx = \dfrac{1}{\sqrt{2\pi}} \int_{-0.5}^{\infty} e^{-x^2/2}\, dx$

   (d) $\dfrac{1}{\sqrt{2\pi} \cdot 100} \displaystyle\int_{-\infty}^{600} \exp\left(-\frac{x - 500}{2 \cdot 100^2}\right) dx = \dfrac{1}{\sqrt{2\pi}} \int_{-\infty}^{1} e^{-x^2/2}\, dx$

   (e) $\dfrac{1}{\sqrt{2\pi} \cdot 100} \displaystyle\int_{450}^{600} \exp\left(-\frac{x - 500}{2 \cdot 100^2}\right) dx = \dfrac{1}{\sqrt{2\pi}} \int_{-0.5}^{1} e^{-x^2/2}\, dx$

9. (a) Since $n(-t) = n(t)$, $A(-z) = \displaystyle\int_{-\infty}^{-z} n(t)\, dt = \int_z^{\infty} n(t)\, dt = \int_{-\infty}^{\infty} n(t)\, dt - \int_{-\infty}^{z} n(t)\, dt = 1 - A(z)$

   (b) $A(-1.2) = 1 - A(1.2) \approx 1 - 0.8849 = 0.1151$

11. In both parts $m = 10$ and $s = 5$.

   (a) $z = (14 - 10)/5 = 0.8$ and $A(0.8) \approx 0.7881$. Thus, $\dfrac{1}{5\sqrt{2\pi}} \displaystyle\int_{-\infty}^{14} \exp\left(-\dfrac{(x-10)^2}{50}\right) \approx 0.7881$.

   (b) $z = (4 - 10)/5 = -1.2$ and $A(-1.2) = 1 - A(1.2) \approx 0.1151$. Thus, $\dfrac{1}{5\sqrt{2\pi}} \displaystyle\int_{-\infty}^{4} \exp\left(-\dfrac{(x-10)^2}{50}\right) \approx$ 0.1151.

13. Looking at the table shows that the top 10% starts at about 1.3 standard deviations above the mean, i.e., at the raw score 630. Similarly, the top 5% starts at about $Z = 1.7$, i.e., at a raw score of 670.

15. (a) The nearest edge of the net should be placed 135 feet from the cannon (i.e., so that the center of the net is 150 feet from the cannon). This position maximizes the probability that the performer will land in the net.

   (b) Missing the net corresponds to a Z-score of magnitude greater than 1.5. Thus, the probability that the performer will miss the landing net is $\displaystyle\int_{-\infty}^{-1.5} n(x)\,dx + \int_{1.5}^{\infty} n(x)\,dx \approx 0.13361$.

17. (a) Since $A''(x) = -\dfrac{x}{\sqrt{2\pi}} e^{-x^2/2}$, the graph of A is concave down over the interval $[0.7, 0.8]$. Therefore, the graph lies *above* the secant line joining $\big(0.7,\, A(0.7)\big)$ and $\big(0.8,\, A(0.8)\big)$. In other words, $A(0.75) > 0.77305$.

   (b) $A(0.75) \approx A(0.7) + 0.05 \cdot A'(0.7) = 0.7580 + 0.05 \cdot 0.31225 \approx 0.7736$

   (c) Since A is concave down over the interval $[0.7, 0.8]$, the line tangent to the graph $y = A(x)$ at $x = 0.7$ lies above the graph at $x = 0.75$. Thus, the estimate in part (b) overestimates $A(0.75)$.

19. The idea is to estimate various areas under the graph, perhaps using the fact that each grid rectangle has area 0.05. Reasonable answers are below; they correspond to probabilities that an observation falls in the given range:

   (a) area about 0.2.   (c) area about 0.875.
   (b) area about 0.4.   (d) area about 0.125.

21. Since $\displaystyle\int_{-\infty}^{\infty} f(x)\,dx = k\pi$, f is a probability density function if $k = 1/\pi$.

23. (a) f is a probability density function because it is a positive function and

$$\int_{-\infty}^{\infty} f(x)\,dx = \int_{0}^{\infty} \lambda e^{-\lambda x}\,dx = -e^{-\lambda x}\Big]_{0}^{\infty} = 1.$$

   (b) $\displaystyle\int_{-\infty}^{\infty} x f(x)\,dx = \lambda \int_{0}^{\infty} x e^{-\lambda x}\,dx = -\dfrac{1+\lambda x}{\lambda} e^{-\lambda x}\Big]_{0}^{\infty} = \dfrac{1}{\lambda}$

10.4 l'Hôpital's Rule: Comparing Rates

1. (a)

x	1	10	100	1000
x^2	1	100	10,000	1×10^6
2^x	2	1024	1.268×10^{30}	1.07×10^{301}
$(3x + 10)^2$	169	1600	96,100	9.0601×10^6

(b) One might guess $\lim\limits_{x \to \infty} \left(\dfrac{x^2}{2^x} \right) = 0$; $\lim\limits_{x \to \infty} \left(\dfrac{x^2}{(3x + 10)^2} \right) = \dfrac{1}{9}$; $\lim\limits_{x \to \infty} \left(\dfrac{2^x}{(3x + 10)^2} \right) = \infty$

(c) Using l'Hôpital's rule:

$$\lim_{x \to \infty} \left(\frac{x^2}{2^x} \right) = \lim_{x \to \infty} \left(\frac{2x}{2^x \ln 2} \right) = \lim_{x \to \infty} \left(\frac{2}{2^x (\ln 2)^2} \right) = 0$$

$$\lim_{x \to \infty} \left(\frac{x^2}{(3x + 10)^2} \right) = \lim_{x \to \infty} \left(\frac{2x}{2(3x + 10) \cdot 3} \right) = \lim_{x \to \infty} \frac{2}{18} = \frac{1}{9}$$

$$\lim_{x \to \infty} \left(\frac{2^x}{(3x + 10)^2} \right) = \lim_{x \to \infty} \left(\frac{2^x \ln 2}{2(3x + 10) \cdot 3} \right) = \lim_{x \to \infty} \left(\frac{2^x (\ln 2)^2}{18} \right) = \infty$$

3. $\lim\limits_{x \to \infty} \left(\dfrac{x^2 + 1}{2x^2 + 3} \right) = \lim\limits_{x \to \infty} \left(\dfrac{2x}{4x} \right) = \dfrac{1}{2}$

5. $\lim\limits_{x \to 0} \left(\dfrac{5x - \sin x}{x} \right) = \lim\limits_{x \to 0} \left(\dfrac{5 - \cos x}{1} \right) = 4$

7. $\lim\limits_{x \to 0} \left(\dfrac{1 - \cos(5x)}{4x + 3x^2} \right) = \lim\limits_{x \to 0} \left(\dfrac{\sin(5x)\ 5}{4 + 6x} \right) = 0$

9. $\lim\limits_{x \to 0} \dfrac{e^x - x - 1}{x^2} = \lim\limits_{x \to 0} \dfrac{e^x - 1}{2x} = \lim\limits_{x \to 0} \dfrac{e^x}{2} = \dfrac{1}{2}$

11. (a) Since $\lim\limits_{x \to 1} f(x) = \lim\limits_{x \to 1} (x^2 - 1) = 0$, $\lim\limits_{x \to 1} \dfrac{f(x)}{x^2 - 1} = \lim\limits_{x \to 1} \dfrac{f'(x)}{2x} = \dfrac{f'(1)}{2} = -1.5$.

(b) Since $\lim\limits_{x \to 2} f(x) = -1.6$, $\lim\limits_{x \to 2^-} (x^2 - 4)^{-1} = -\infty$, and $\lim\limits_{x \to 2^+} (x^2 - 4)^{-1} = \infty$, $\lim\limits_{x \to 2} \dfrac{f(x)}{x^2 - 4}$ does not exist.

(c) Since $\lim\limits_{x \to 1} f(x) = 0$ and $\lim\limits_{x \to 1} f(-2x) = 0$, $\lim\limits_{x \to 1} \dfrac{f(x)}{f(-2x)} = \lim\limits_{x \to 1} \dfrac{f'(x)}{-2f'(-2x)} = \dfrac{f'(1)}{-2f'(-2)} \approx 0.3125$.

(d) Since $\lim\limits_{x \to 1} f(x - 3) = 0$ and $\lim\limits_{x \to 1} f(x + 3) = 3.6$, $\lim\limits_{x \to 1} \dfrac{f(x - 3)}{f(x + 3)} = 0$.

13. $\lim\limits_{x \to 0} x \cot x = \lim\limits_{x \to 0} \left(\dfrac{x}{\tan x} \right) = \lim\limits_{x \to 0} \left(\dfrac{1}{\sec^2 x} \right) = 1$

15. $\lim\limits_{x \to 0^+} \left(\dfrac{\sin x}{x + \sqrt{x}} \right) = \lim\limits_{x \to 0^+} \left(\dfrac{\cos x}{1 + \frac{1}{2\sqrt{x}}} \right) \to \dfrac{1}{\infty} = 0$

17. $\lim\limits_{x \to 0} \dfrac{\tan(3x)}{\ln(1 + x)} = \lim\limits_{x \to 0} \dfrac{3 \sec^2(3x)}{(1 + x)^{-1}} = 3$

19. $\displaystyle\lim_{x\to 0}\frac{\arctan(2x)}{\sin(3x)} = \lim_{x\to 0}\frac{2/(1+x^2)}{3\cos(3x)} = \frac{2}{3}$

21. $\displaystyle\lim_{x\to 0}\frac{\sin^2 x}{\cos(3x)-1} = \lim_{x\to 0}\frac{2\sin x\cos x}{-3\sin(3x)} = \lim_{x\to 0}\frac{2\cos^2 x - 2\sin^2 x}{-9\cos(3x)} = -\frac{2}{9}$

23. $\displaystyle\lim_{x\to 0}\frac{1-x-e^{-x}}{1-\cos x} = \lim_{x\to 0}\frac{e^{-x}-1}{\sin x} = \lim_{x\to 0}\frac{-e^{-x}}{\cos x} = -1$

25. $\displaystyle\lim_{x\to\infty} x^2 e^{-x^2/2} = \lim_{x\to\infty}\frac{x^2}{e^{x^2/2}} = \lim_{x\to\infty}\frac{2}{e^{x^2/2}} = 0$

27. $\displaystyle\lim_{x\to 0}\frac{1-x^2-e^{-x^2}}{x^4} = \lim_{x\to 0}\frac{e^{-x^2}-1}{2x^2} = \lim_{x\to 0}\frac{-e^{-x^2}}{2} = -\frac{1}{2}$

29. $\displaystyle\lim_{x\to\pi/2}\frac{\ln(\sin x)}{(x-\pi/2)^2} = \lim_{x\to\pi/2}\frac{\cos x}{2(x-\pi/2)\sin x} = \lim_{x\to\pi/2}\frac{-\sin x}{2\sin x + 2(x-\pi/2)\cos x} = -\frac{1}{2}$

31. $\displaystyle\lim_{x\to\infty} x\sin(1/x) = \lim_{x\to\infty}\frac{\sin(1/x)}{1/x} = \lim_{x\to\infty}\cos(1/x) = 1$

33. $\displaystyle\lim_{x\to 0^+}\sin x\,\ln(\sin x) = \lim_{x\to 0^+}\frac{\ln(\sin x)}{1/\sin x} = \lim_{x\to 0^+} -\sin x = 0$

35. $\displaystyle\lim_{x\to\pi/2}\left(\frac{\pi}{2}-x\right)\tan x = \lim_{x\to\pi/2}\frac{\pi/2-x}{\cot x} = \lim_{x\to\pi/2}\sin^2 x = 1$

37. $\displaystyle\lim_{x\to\infty}\frac{\int_1^x\sqrt{1+e^{-3t}}\,dt}{x} = \lim_{x\to\infty}\frac{\sqrt{1+e^{-3x}}}{1} = 1$

[NOTE: The improper integral in the numerator diverges to ∞ because $\sqrt{1+e^{-3t}} > 1$ for all $t \geq 1$.]

39. $\displaystyle\lim_{x\to\infty} e^{x^2}\int_0^x e^{-t^2}\,dt = \infty$ since $\displaystyle\lim_{x\to\infty}\int_0^x e^{-t^2}\,dt = \sqrt{\pi}/2$ and $\displaystyle\lim_{x\to\infty} e^{x^2} = \infty$.

41. $f(k) = \begin{cases} \dfrac{3k-6}{k-2} & \text{when } k \neq 2 \\ -1 & \text{when } k = 2 \end{cases}$

43. $\displaystyle\lim_{x\to 0}\frac{f(x)}{\sin(2x)} = \lim_{x\to 0}\frac{f'(x)}{2\cos(2x)} = \frac{f'(0)}{2} = 5$ so $f'(0) = 10$.

45. $\displaystyle\lim_{x\to\infty}\frac{g(1/x)}{f(1/x)} = \lim_{u\to 0^+}\frac{g(u)}{f(u)} = -\frac{1}{2}$

47. (a) $\displaystyle\lim_{x\to 0}\sin x = \lim_{x\to 0} x = 0$ but $\displaystyle\lim_{x\to 0}\cos x = 1$.

(b) $\displaystyle\lim_{x\to 0^+}\frac{\cos x}{x} = \infty$

(c) **Diverges** since $\cos x > 1/2$ when $0 \leq x \leq 1 \implies \displaystyle\int_0^1\frac{\cos x}{x}\,dx > \frac{1}{2}\int_0^1\frac{dx}{x} = \infty$.

(d) $\displaystyle\lim_{x\to 0^+}\frac{\sin x}{x} = 1$ so the integrand is bounded over the entire interval of integration.

(e) $\displaystyle\int_0^\infty\sin\left(e^{-x}\right)dx = \lim_{t\to\infty}\int_0^t\sin\left(e^{-x}\right)dx = \lim_{t\to\infty} -\int_1^{e^{-t}}\sin u\,\frac{du}{u} = \int_0^1\frac{\sin u}{u}\,du$. The desired inequalities follow from the fact that $0.8 < \dfrac{\sin x}{x} < 1$ when $0 < x < 1$.

49. No. Using L'Hôpital's rule, $\lim\limits_{x \to 0} \dfrac{\arctan x}{\sqrt[3]{x}} = \lim\limits_{x \to 0} \dfrac{(1+x^2)^{-1}}{\frac{1}{3}x^{-2/3}} = \lim\limits_{x \to 0} \dfrac{3x^{2/3}}{1+x^2} = 0$. Therefore, since the integrand is finite throughout the interval of integration, the integral is not improper.

51. (a) $\left| \displaystyle\int_0^\infty \dfrac{\sin x}{x^{3/2}}\, dx \right| = \left| \displaystyle\int_0^1 \dfrac{\sin x}{x^{3/2}}\, dx + \int_1^\infty \dfrac{\sin x}{x^{3/2}}\, dx \right| \le \left| \displaystyle\int_0^1 \dfrac{\sin x}{x^{3/2}}\, dx \right| + \left| \displaystyle\int_1^\infty \dfrac{\sin x}{x^{3/2}}\, dx \right|$

$$\le \int_0^1 \dfrac{x}{x^{3/2}}\, dx + \int_1^\infty \dfrac{1}{x^{3/2}}\, dx = \int_0^1 \dfrac{1}{x^{1/2}}\, dx + \int_1^\infty \dfrac{1}{x^{3/2}}\, dx = 2 + 2$$

Therefore, Theorem 2 (p. 195) asserts that the improper integral $\displaystyle\int_0^\infty f(x)\, dx$ converges.

(b) $\displaystyle\int_0^\alpha \dfrac{\sin x}{x^{3/2}}\, dx \le \int_0^\alpha \dfrac{dx}{\sqrt{x}} = 2\sqrt{\alpha} \le 0.0005$ when $\alpha \le 6.25 \times 10^{-8}$.

$$\left| \int_\beta^\infty \dfrac{\sin x}{x^{3/2}}\, dx \right| \le \int_\beta^\infty \dfrac{dx}{x^{3/2}} = \dfrac{2}{\sqrt{\beta}} \le 0.0005 \text{ when } \beta \ge 1.6 \times 10^7.$$

Therefore, $\displaystyle\int_{6.25\times 10^{-8}}^{1.6\times 10^7} \dfrac{\sin x}{x^{3/2}}\, dx$ approximates $\displaystyle\int_0^\infty \dfrac{\sin x}{x^{3/2}}\, dx$ within 0.001.

53. $\lim\limits_{x \to 0} \dfrac{\tan x}{x} = \lim\limits_{x \to 0} \sec^2 x = 1$ so $\lim\limits_{x \to 0} \cos\left(\dfrac{\tan x}{x}\right) = \cos 1$

55. $\lim\limits_{x \to 0^+} 2x \ln x = 0$ so $\lim\limits_{x \to 0^+} x^{2x} = e^0 = 1$

57. $\lim\limits_{x \to 0} \dfrac{\ln(1+x)}{x} = \lim\limits_{x \to 0} \dfrac{1}{1+x} = 1$ so $\lim\limits_{x \to \infty} (1+x)^{1/x} = e^1 = e$.

59. $\lim\limits_{x \to 1} (\ln x)^{\sin x} = 0$ since $\lim\limits_{x \to 1} \ln x = 0$ and $\lim\limits_{x \to 1} \sin x = \sin 1 \approx 0.84147$.

61. $\displaystyle\int_1^\infty f'(x)\, dx = \lim\limits_{t \to \infty} f(t) - f(1)$. Thus, the improper integral converges if (and only if) $\lim\limits_{t \to \infty} f(t)$ is a finite number.

Since $|f(x)| \le e^{-x} \ln x$, $\lim\limits_{t \to \infty} |f(t)| \le \lim\limits_{t \to \infty} e^{-t} \ln t = 0$ (see problem 3d above). Thus, $\lim\limits_{t \to \infty} f(t) = 0$.

Finally, therefore, $\displaystyle\int_1^\infty f'(x)\, dx = -f(1)$.

63. (a) $\lim\limits_{x \to a} \dfrac{f(x) - P_2(x)}{x - a} = \lim\limits_{x \to a} \left(f'(x) - P_2'(x) \right) = f'(a) - P_2'(a) = 0$

(b) $\lim\limits_{x \to a} \dfrac{f(x) - P_2(x)}{(x - a)^2} = \lim\limits_{x \to a} \dfrac{f'(x) - P_2'(x)}{2(x - a)} = \lim\limits_{x \to a} \dfrac{f''(x) - P_2''(x)}{2} = 0$

(c) $\lim\limits_{x \to a} \dfrac{f(x) - q(x)}{(x - a)^2} = \lim\limits_{x \to a} \dfrac{f'(x) - q'(x)}{2(x - a)} = \lim\limits_{x \to a} \dfrac{f''(x) - q''(x)}{2} \ne 0$ since $q''(a) \ne f''(a)$. (If $q''(a) = f''(a)$, then $q(x) = P_2(x)$ would be true.)

(d) The graph of P_2 is more like the graph of f near $x = a$ than the graph of any other quadratic polynomial.

(e) Since $f(a) = P_n(a)$ and the first n derivatives of f and P_n are equal at $x = a$, l'Hôpital's rule implies that $\lim\limits_{x \to a} \dfrac{f(x) - P_n(x)}{(x - a)^n} = 0$. No other n^{th}-order polynomial has this property since any n^{th}-order polynomial that is not P_n will not have the same derivative values at $x = a$ as f does.

65. Let $t = \ln(1/x)$. Then $e^{-t} = x$ and $dt = -(1/x)\,dx$ so

$$\int_0^\infty t^z e^{-t}\,dt = \lim_{s \to \infty} \int_0^s t^z e^{-t}\,dt = \lim_{w \to 0^+} -\int_1^w \big(\ln(1/x)\big)^z\,dx = \int_0^1 \big(\ln(1/x)\big)^z\,dx.$$

11.1   Sequences and Their Limits

1. $a_k = (-1/3)^k$ for $k = 0, 1, 2, \ldots$

3. $a_k = k/2^k$ for $k = 1, 2, 3 \ldots$

5. $\lim\limits_{k \to \infty} a_k$ does not exist

7. $\lim\limits_{k \to \infty} a_k = \infty$

9. $\lim\limits_{k \to \infty} a_k = \infty$

11. $\lim\limits_{k \to \infty} a_k = \infty$

13. $\lim\limits_{k \to \infty} a_k$ does not exist

15. $\lim\limits_{k \to \infty} a_k = 1$

17. $\lim\limits_{k \to \infty} a_k$ does not exist

19. $\lim\limits_{k \to \infty} a_k = \sqrt{2}$

21. $\lim\limits_{n \to \infty} a_n = \lim\limits_{n \to \infty} \dfrac{n+2}{n^3+4} = \lim\limits_{n \to \infty} \dfrac{1}{3n^2} = 0$

23. $\lim\limits_{j \to \infty} b_j = \lim\limits_{j \to \infty} \dfrac{\ln j}{\sqrt[3]{j}} = \lim\limits_{j \to \infty} \dfrac{3j^{2/3}}{j} = \lim\limits_{j \to \infty} \dfrac{3}{\sqrt[3]{j}} = 0$

25. $\lim\limits_{n \to \infty} d_n = \lim\limits_{n \to \infty} n \sin(1/n) = \lim\limits_{n \to \infty} \dfrac{\sin\left(n^{-1}\right)}{n^{-1}} = \lim\limits_{n \to \infty} \cos\left(n^{-1}\right) = 1$

27. $\lim\limits_{n \to \infty} \dfrac{\ln x}{n} = 0 \implies \lim\limits_{n \to \infty} x^{1/n} = 1$ for all $x > 0$.

29. Let $a_n = \left(1 - \dfrac{1}{2n}\right)^n$. Then $\ln(a_n) = n \ln\left(1 - \dfrac{1}{2n}\right) = \dfrac{\ln\left(1 - \frac{1}{2n}\right)}{\frac{1}{n}}$ so, using l'Hôpital's rule, $\lim\limits_{n \to \infty} \ln(a_n) = -1/2$. It follows that $\lim\limits_{n \to \infty} a_n = e^{-1/2}$.

31. $\lim\limits_{k \to \infty} a_k = 0$

33. $\lim\limits_{k \to \infty} a_k = 0$

35. $\lim\limits_{k \to \infty} a_k = 0$

37. $\lim\limits_{k \to \infty} a_k = 1/2$

39. (a) $a_k = (-1)^k/k$. $\lim\limits_{k \to \infty} a_k = 0$, but $\{a_k\}$ is neither an increasing nor a decreasing sequence.

   (b) $a_k = \cos k$. $|a_k| \leq 1$ for all integers $k \geq 1$, but $lim_{k \to \infty} a_k$ does not exist.

   (c) $a_k = k$. $a_k < a_{k+1}$ for all integers $k \geq 1$, but $lim_{k \to \infty} a_k = \infty$.

   (d) $a_k = -k$. $a_{k+1} < a_k$ for all integers $k \geq 1$, and $lim_{k \to \infty} a_k = -\infty$.

   (e) $a_k = e^{-k}$. $a_{k+1} < a_k$ for all integers $k \geq 1$, and $lim_{k \to \infty} a_k = 0$.

(f) $a_k = (-1)^k k$. $a_k < a_{k+1}$ when k is an odd integer, but $a_{k+1} < a_k$ when k is an even integer. Thus, $\{ a_k \}$ is not a monotone sequence. Also, $\lim\limits_{k \to \infty} |a_k| = \infty$ so the sequence is not bounded.

41. (a) $a_{n+1} = \displaystyle\sum_{k=1}^{n+1} \frac{1}{(n+1)+k} = \sum_{k=2}^{n+2} \frac{1}{n+k} = \sum_{k=2}^{n} \frac{1}{n+k} + \frac{1}{2n+1} + \frac{1}{2n+2}$

$> \displaystyle\sum_{k=2}^{n} \frac{1}{n+k} + \frac{2}{2n+2} = \sum_{k=1}^{n} \frac{1}{n+k} = a_n$

(b) $a_n = \displaystyle\sum_{k=1}^{n} \frac{1}{n+k} \le \sum_{k=1}^{n} \frac{1}{n+1} = \frac{n}{n+1} < 1$ when $n \ge 1$.

(c) Parts (a) and (b) imply that the sequence $\{ a_n \}$ is monotonically increasing and bounded above. Therefore, $\lim\limits_{n \to \infty} a_n$ exists.

(d) Part (a) implies that the sequence $\{ a_n \}$ is montonically increasing. Therefore, $\lim\limits_{n \to \infty} a_n > a_k$ for any integer $k \ge 1$. Since $a_1 = 1/2$, $\lim\limits_{n \to \infty} a_n > 1/2$.

(e) $\lim\limits_{n \to \infty} a_n = \displaystyle\int_0^1 \frac{dx}{1+x} = \ln 2$

43. (a) First, note that $\ln(\sqrt[n]{n!}) = \ln(n!)^{1/n} = \frac{1}{n} \ln(n!) = \frac{1}{n} \displaystyle\sum_{k=1}^{n} \ln k$. Also, $n = (n^n)^{1/n}$ so

$\ln n = \frac{1}{n} \ln(n^n) = \frac{1}{n} \displaystyle\sum_{k=1}^{n} \ln n$. Therefore, $\ln a_n = \dfrac{\sqrt[n]{n!}}{n} = \ln\left(\sqrt[n]{n!}\right) - \ln n = \frac{1}{n} \displaystyle\sum_{k=1}^{n} \ln k - \frac{1}{n} \sum_{k=1}^{n} \ln n$.

(b) The right sum approximation to $\displaystyle\int_0^1 \ln x \, dx$ is

$$R_n = \frac{1}{n} \sum_{k=1}^{n} \ln\left(\frac{k}{n}\right) = \frac{1}{n} \sum_{k=1}^{n} \ln k - \frac{1}{n} \sum_{k=1}^{n} \ln n.$$

(c) The result in part (b) implies that $\lim\limits_{n \to \infty} \ln a_n = \displaystyle\int_0^1 \ln x \, dx = x \ln x - x \Big]_0^1 = -1$. Therefore, $\lim\limits_{n \to \infty} a_n = e^{-1}$.

45. Observe that $\dfrac{n+3}{2n+1} \le \dfrac{6}{7}$ for all $n \ge 3$. Now,

$$\frac{a_{n+1}}{a_3} = \frac{a_{n+1}}{a_n} \cdot \frac{a_n}{a_{n-1}} \cdots \frac{a_4}{a_3} \le \left(\frac{6}{7}\right)^{n-2} \implies a_{n+1} \le (6/7)^{n-2} a_3 \implies \lim_{n \to \infty} a_n = 0.$$

47. $\lim\limits_{k \to \infty} a_k$ exists only when $x \le 0$ because $e^x > 1$ when $x > 0$. $\lim\limits_{k \to \infty} e^k x = \infty$. When $x < 0$, $\lim\limits_{k \to \infty} a_k = 0$. When $x = 0$, $\lim\limits_{k \to \infty} a_k = 1$.

49. $\lim\limits_{k \to \infty} a_k = 0$ when $-\sin 1 < x < \sin 1$ and $\lim\limits_{k \to \infty} a_k = 1$ when $x = \sin 1$

51. (a) $a_1 \approx 0.5403$, $a_2 \approx -0.2248$, $a_3 \approx 0.2226$, $a_4 \approx -0.1455$

(b) Yes. $|a_n| \le 1$ since $|\cos x| \le 1$ for all x.

(c) No. The terms of the sequence change sign.

(d) Yes. $|a_n| \le 1$.

(e) Yes. $|a_{n+1}| < a_n$

53. (a) $a_1 = 1/2$; $a_2 = 3/8$; $a_5 = 945/3840 = 63/256$

(b) Since $0 < a_{n+1} < a_n$, the sequence is bounded below and monotonically decreasing. Therefore, $\lim\limits_{n \to \infty} a_n$ exists.

55. No. The terms of the sequence are 1 and 0 alternately.

57. (a) Since $0 < a_{n+1} < a_n$, the sequence is bounded below and monotone decreasing. Therefore, it must have a limit.

(b) $\lim\limits_{n \to \infty} a_n = \lim\limits_{n \to \infty} 1/n = 0$

59. $a_{n+1} = x^{1/2^n}$ so $\lim\limits_{n \to \infty} a_n = 1$ for all $x > 0$ by the previous problem. When $x = 0$, $\lim\limits_{n \to \infty} a_n = 0$. Thus, $\lim\limits_{n \to \infty} a_n$ exists for all $x \geq 0$.

61.

(a) Dividing both sides of the equation $F_{n+1} = F_n + F_{n-1}$ by F_n leads to the equation

$$\frac{F_{n+1}}{F_n} = 1 + \frac{F_{n-1}}{F_n} = 1 + \frac{1}{F_n/F_{n-1}}.$$

Now, taking the limit as $n \to \infty$, we obtain the equation $L = 1 + 1/L$ which is equivalent to $L^2 = L + 1$.

(b) Since $x = xF_1 + F_0 = x + 1$, the equation holds when $n = 1$. Therefore, assume that the equation is true when $n = N$. We must show that the equation holds when $n = N + 1$:

$$x^{N+1} = x\,x^N = x\,(xF_N + F_{N-1})$$
$$= (x + 1)F_N + xF_{N-1} = x\,(F_N + F_{N-1}) + F_N$$
$$= xF_{N+1} + F_N$$

(c) The numbers r_1 and r_2 are solutions of the equation $x^2 = x + 1$. Therefore, part (b) implies that $r_1^n - r_2^n = (r_1 F_n + F_{n-1}) - (r_2 F_n + F_{n-1}) = (r_1 - r_2) F_n = \sqrt{5} F_n$.

(d) Since $r_2 < r_1$,

$$\lim_{n \to \infty} \frac{F_{n+1}}{F_n} = \lim_{n \to \infty} \frac{r_1^{n+1} - r_2^{n+1}}{r_1^n - r_2^n} = \lim_{n \to \infty} \frac{r_1 - \left(\frac{r_2}{r_1}\right)^n r_2}{1 - \left(\frac{r_2}{r_1}\right)^n} = r_1 = \frac{1 + \sqrt{5}}{2}.$$

11.2   Infinite Series, Convergence, and Divergence

1. (a) $S_5 = 63$; yes

   (b) $S_5 = \dfrac{101010101010}{100000000000}$; yes

   (c) When $r = 1$, $S_n = (n+1)a$ but the right-hand side of the expression is undefined (because of the zero in the denominator).

   (d) $S_n - rS_n = (a + ar + ar^2 + \cdots + ar^n) - r(a + ar + ar^2 + \cdots + ar^n) = a - ar^{n+1} = a(1 - r^{n+1})$

   (e) $S_n - rS_n = (1-r)S_n = a(1 - r^{n+1})$. The desired result is now obtained by dividing through by $1 - r$.

   (f) The sum is a geometric series with $a = 3$, $r = 2$, and $n = 10$. Thus, $3 + 6 + 12 + \cdots + 3072 = 6141$.

3. (a) $a_1 = 1$; $a_2 = 1/4$; $a_5 = 1/25$; $a_{10} = 1/100$

   (b) $S_1 = 1$; $S_2 = 5/4$; $S_5 = 5269/3600 \approx 1.46361$; $S_{10} = 1968329/1270080 \approx 1.54977$

   (c) Yes. S_n is an increasing sequence because the terms of the series are all positive.

   (d) $R_1 = \pi^2/6 - S_1 \approx 0.64493$; $R_2 = \pi^2/6 - S_2 \approx 0.39493$; $R_5 = \pi^2/6 - S_5 \approx 0.18132$; $R_{10} = \pi^2/6 - S_{10} \approx 0.095166$

   (e) Yes because $R_n = \pi^2/6 - S_n$ and $S_{n+1} > S_n$.

   (f) $R_{20} \approx 0.048771$. Since $0 < R_{n+1} < R_n$ for all $n \geq 1$, $0 < R_n < 0.05$ for all $n \geq 20$.

   (g) $\lim\limits_{n \to \infty} R_n = 0$

5. (a) $a_1 = -4/5 = -0.8$; $a_2 = 16/25 = 0.64$; $a_5 = 1024/3125 = -0.32768$; $a_{10} = 1048576/9765625 = 0.1073741824$

   (b) $S_1 = 1/5 = 0.2$; $S_2 = 21/25 = 0.84$; $S_5 = 1281/3125 = 0.40992$; $S_{10} = 5891381/9765625 \approx 0.60328$

   (c) $\displaystyle\sum_{k=0}^{\infty} (-0.8)^k = \dfrac{1}{1-(-0.8)} = \dfrac{5}{9} \approx 0.5555555555$

   (d) $R_1 = 16/45 \approx 0.35556$; $R_2 = -64/225 \approx -0.28444$; $R_5 = 4096/28125 \approx 0.14564$; $R_{10} = -4194304/87890625 \approx -0.047722$

   (e) No — $S_{2m+1} < S_{2m} < S_{2m-1}$ for $m = 1, 2, 3, \ldots$ (the partial sums are go up and down in value)

   (f) $R_n = \displaystyle\sum_{k=n+1}^{\infty} (-0.8)^k = \dfrac{5}{9}(-0.8)^{n+1}$ so the terms of the sequence are alternately positive and negative.

   (g) $|R_{n+1}| = \left| \displaystyle\sum_{k=n+2}^{\infty} (-0.8)^k \right| = \left| -0.8 \displaystyle\sum_{k=n+1}^{\infty} (-0.8)^k \right| = 0.8\,|R_n| < |R_n|$ (i.e., the sequence is decreasing).

   (h) $\lim\limits_{n \to \infty} R_n = 0$.

7. (a) $S_1 = 8/15 \approx 0.53333$; $S_2 = 103/165 \approx 0.62424$; $S_5 = 2626616/3892119 \approx 0.67486$; $S_{10} = \dfrac{2923789575132170780644253}{4319382292184626122988455} \approx 0.67690$

   (b) The sequence of partial sums is increasing because each term of the series is a positive number (i.e., $S_{n+1} = S_n + a_{n+1} > S_n$ because $a_{n+1} > 0$).

   Since $a_j < 3^{-j}$ for all $j \geq 0$, $S_n = \displaystyle\sum_{j=0}^{n} a_j < \sum_{j=0}^{n} 3^{-j} = \dfrac{1}{2}\left(3 - 3^{-n}\right) \leq \dfrac{3}{2}$ for all $n \geq 0$. Thus, each term in the sequence of partial sums is bounded above by $3/2$.

   (c) Since the sequence of partial sums is increasing and bounded above, it must converge. This implies that the infinite series converges.

9. $\dfrac{1}{16} + \dfrac{1}{32} + \dfrac{1}{64} + \dfrac{1}{128} + \cdots + \dfrac{1}{2^{i+4}} + \cdots = \dfrac{1}{2^4} \displaystyle\sum_{i=0}^{\infty} \dfrac{1}{2^i} = \dfrac{1}{2^4} \cdot 2 = \dfrac{1}{8}$

11. $\displaystyle\sum_{n=0}^{\infty} e^{-n} = \frac{e}{e-1}$

13. $\displaystyle\sum_{m=1}^{\infty} (\arctan 1)^m = \sum_{m=1}^{\infty} (\pi/4)^m = \sum_{m=0}^{\infty} (\pi/4)^m - 1 = \frac{1}{1-\pi/4} - 1 = \frac{\pi}{4-\pi}$

15. $\displaystyle\sum_{j=5}^{\infty} \left(-\frac{1}{2}\right)^j = \sum_{j=0}^{\infty} \left(-\frac{1}{2}\right)^j - \sum_{j=0}^{4} \left(-\frac{1}{2}\right)^j = \frac{1}{1-(-1/2)} - \frac{1-(-1/2)^5}{1-(-1/2)} = -\frac{1}{48} \approx -0.020833$

17. Since $1/(2 + \sin k) \geq 1$ for all $k \geq 1$, the series diverges by the nth-term test.

19. $S_n = \arctan(n+1) - \arctan(0) = \arctan(n+1)$. Since $\displaystyle\lim_{n \to \infty} S_n = \frac{\pi}{2}$, the series converges to $\frac{\pi}{2}$.

21. $\displaystyle\sum_{k=1}^{\infty} \frac{2}{k^2+k} = \sum_{k=1}^{\infty} \frac{2}{k} - \frac{2}{k+1}$ so $S_n = 2 - \frac{2}{n+1}$. Since $\displaystyle\lim_{n \to \infty} S_n = 2$, the series converges to 2.

23. $S_n = 1 + \dfrac{1}{\sqrt{2}} - \dfrac{1}{\sqrt{n+1}} - \dfrac{1}{\sqrt{n+2}}$ when $n \geq 1$. Since $\displaystyle\lim_{n \to \infty} S_n = 1 + 1/\sqrt{2}$, the series converges to $1 + 1/\sqrt{2}$.

25. $S_n = \left(1 + (-1)^n\right)/2$. Since $\displaystyle\lim_{n \to \infty} S_n$ doesn't exist, the series diverges.

27. (a) $\dfrac{1}{4} + \dfrac{1}{16} + \dfrac{1}{36} + \dfrac{1}{100} + \cdots = \dfrac{1}{4}\left(1 + \dfrac{1}{4} + \dfrac{1}{9} + \cdots\right) = \dfrac{\pi^2}{24}$.

   (b) $\displaystyle\sum_{k=0}^{\infty} \frac{1}{(2k+1)^2} = \sum_{k=1}^{\infty} \frac{1}{k^2} - \frac{1}{4} \sum_{k=1}^{\infty} \frac{1}{k^2} = \frac{\pi^2}{6} - \frac{\pi^2}{24} = \frac{\pi^2}{8}$.

   (c) $\displaystyle\sum_{m=1}^{\infty} \frac{(-1)^{m+1}}{m^2} = \sum_{m=1}^{\infty} \frac{1}{m^2} - \frac{1}{2} \sum_{m=1}^{\infty} \frac{1}{m^2} = \frac{\pi^2}{12}$.

   Alternatively, $\displaystyle\sum_{m=1}^{\infty} \frac{(-1)^{m+1}}{m^2} = \sum_{k=0}^{\infty} \frac{1}{(2k+1)^2} - \sum_{j=1}^{\infty} \frac{1}{(2j)^2} = \frac{\pi^2}{8} - \frac{\pi^2}{24} = \frac{\pi^2}{12}$.

29. The series converges for all values of x such that $-1 < x < 1$. $\displaystyle\sum_{k=0}^{\infty} x^k = \frac{1}{1-x}$.

31. Since $\displaystyle\sum_{j=5}^{\infty} x^{2j} = x^{10} \sum_{j=0}^{\infty} \left(x^2\right)^j$, the series converges when $-1 < x^2 < 1$. Thus, the series converges to $\dfrac{x^{10}}{1-x^2}$ for all values of x such that $-1 < x < 1$.

33. The series converges when $|1 + x| < 1$. Thus, the series converges for all values of x such that $-2 < x < 0$.
$\displaystyle\sum_{n=3}^{\infty} (1+x)^n = (1+x)^3 \sum_{n=0}^{\infty} x^n = -\frac{(1+x)^3}{x}$.

35. Since $|\sin x|/2 < 1$ for all x, the series converges to $\dfrac{2}{2 - \sin x}$ for all x.

37. The series converges when $|\arctan x| < 1$. Thus, the series converges when $|x| < \pi/4$. For these values of x,
$\displaystyle\sum_{n=0}^{\infty} (\arctan x)^n = \frac{1}{1 - \arctan x}$.

39. $a_{n+1} = 4 - \displaystyle\sum_{k=1}^{n}(1/2)^k \implies \lim_{n\to\infty} a_n = 3$

41. The series $\displaystyle\sum_{k=0}^{\infty} \frac{3}{10}\left(-\frac{1}{2}\right)^k$ converges to $1/5$.

43. Diverges by the nth term test.

45. Diverges by the nth term test: $\displaystyle\lim_{n\to\infty}\frac{n+1}{2n+1} = \frac{1}{2} \neq 0$.

47. The partial sums of the series are $S_N = \displaystyle\sum_{n=2}^{N}\frac{2}{n^2-1} = \sum_{n=2}^{N}\left(\frac{1}{n-1}-\frac{1}{n+1}\right) = 1 + \frac{1}{2} - \frac{1}{N} - \frac{1}{N+1}$. Therefore, the series converges to $3/2$.

49. $\displaystyle\sum_{m=2}^{\infty}\frac{1}{(\ln 3)^m} = \frac{1}{(\ln 3)^2}\frac{1}{1-(1/\ln 3)} = \frac{1}{(\ln 3)(\ln 3 - 1)}$

51. $\displaystyle\sum_{k=1}^{\infty}\left(\int_{k}^{k+1}\frac{dx}{x^2}\right) = \sum_{k=1}^{\infty}\frac{1}{k(k+1)} = \lim_{n\to\infty}\left(1-\frac{1}{n+1}\right) = 1$

53. Diverges by the nth term test: $\displaystyle\lim_{n\to\infty}\left(1+\frac{1}{n}\right)^n = e \neq 0$.

55. Diverges by the nth term test: $\displaystyle\lim_{n\to\infty}\frac{\ln n}{\ln(3+n^2)} = \frac{1}{2} \neq 0$.

57. (a) $\displaystyle\lim_{k\to\infty}\frac{1}{\sqrt{k}} = 0$

   (b) Since $1/\sqrt{n} \leq 1/\sqrt{k}$ when $1 \leq k \leq n$, $\displaystyle\sum_{k=1}^{n}\frac{1}{\sqrt{k}} \geq \sum_{k=1}^{n}\frac{1}{\sqrt{n}} = \frac{n}{\sqrt{n}} = \sqrt{n}$.

   (c) It follows from part (a) that $\displaystyle\lim_{n\to\infty} S_n = \infty$.

59. The series $\displaystyle\sum_{k=1}^{\infty} a_k$ converges because its partial sums are bounded above (by 100) and increasing (the terms of the series are positive). Therefore, Theorem 5 (p. 230) implies that $\displaystyle\lim_{k\to\infty} a_k = 0$.

61. (a) $\displaystyle\lim_{n\to\infty} S_n = \ln 2$.

   (b) Yes. Since the sequence of partial sums has a limit, the series converges.

   (c) For any $n \geq 1$,

$$b_n = S_n - S_{n-1} = \ln\left(\frac{2n+3}{n+1}\right) - \ln\left(\frac{2n+1}{n}\right)$$

$$= \ln\left(\frac{2n+3}{2n+1}\frac{n}{n+1}\right) = \ln\left(\frac{2n^2+3n}{2n^2+3n+1}\right) < 0.$$

63. (a) $a_k = (-1)^{k+1}$

   (b) By definition, the infinite series $\displaystyle\sum_{k=1}^{\infty} a_k$ converges if the sequence of its partial sums has a limit—that is, the series converges if $\displaystyle\lim_{n\to\infty} S_n$ exists. Since $a_k \geq 0$, S_n is an increasing sequence bounded above by 100 so Theorem 3 implies that this sequence has a limit and, therefore, that the infinite series converges.

(c) $\displaystyle\sum_{k=1}^{\infty} a_k = \sum_{k=1}^{10^6-1} a_k + \sum_{k=10^6}^{\infty} a_k.$ $\displaystyle\sum_{k=1}^{10^6-1} a_k$ is a finite sum, so it is a real number. The infinite series $\displaystyle\sum_{k=10^6}^{\infty} a_k$ converges because the sequence of its partial sums is increasing and bounded above. Thus, the original infinite series converges.

65. The series defining S doesn't converge.

67.

69. (c) $\displaystyle a_k = \sum_{j=1}^{2^{k-1}} \frac{1}{2^{k-1}+j} \geq \sum_{j=1}^{2^{k-1}} \frac{1}{2^{k-1}+2^{k-1}} = \frac{1}{2}$

(d) $\displaystyle H_{2^n} = 1 + \sum_{k=1}^{n} a_k \geq 1 + \sum_{k=1}^{n} \frac{1}{2} = 1 + n/2 \implies \lim_{n\to\infty} H_{2^n} = \infty$

71. (a) Since $\sin 0 = \tan 0 = 0$, and $\cos x \leq 1 \leq \sec^2 x$ when $0 \leq x < \pi/2$ and the racetrack principle implies that $\sin x \leq x \leq \tan x$ when $0 \leq x < \pi/2$.

(b) When $0 < x < \pi/2$, $0 < \sin x \leq x \leq \tan x$ so $0 < \sin^2 x \leq x^2 \leq \tan^2 x$. Therefore, $\csc^2 x \geq 1/x^2 \geq \cot^2 x$ when $0 < x < \pi/2$. The desired result follows from the trigonometric identity $\csc^2 x = 1 + \cot^2 x$.

(c) If k is an integer such that $1 \leq k \leq n$, then $x_k = k\pi/(2n+1)$ satisfies the inequality $0 < x_k < \pi/2$. Therefore, part (b) implies

$$\sum_{k=1}^{n} \cot^2 x_k \leq \frac{1}{x_k^2} \leq \sum_{k=1}^{n}\left(1 + \cot^2 x_k\right).$$

Observing that $\sum_{k=1}^{n} 1 = n$ and inserting the definition of x_k, we obtain the desired result:

$$\sum_{k=1}^{n} \cot^2\left(\frac{k\pi}{2n+1}\right) \leq \frac{(2n+1)^2}{\pi^2} \sum_{k=1}^{n} \frac{1}{k^2} \leq n + \sum_{k=1}^{n} \cot^2\left(\frac{k\pi}{2n+1}\right).$$

(d) Let $\displaystyle S_n = \sum_{k=1}^{n} \frac{1}{k^2}$. After multiplication by $\pi^2/(2n+1)^2$, the given inequality becomes

$$\frac{\pi^2 n(2n-1)}{3(2n+1)^2} \leq S_n \leq \frac{\pi^2 n}{(2n+1)^2} + \frac{\pi^2 n(2n-1)}{3(2n+1)^2}.$$

Evaluating the limit of each term of this inequality as $n \to \infty$ leads to the result

$$\frac{\pi^2}{6} \leq \sum_{k=1}^{\infty} \frac{1}{k^2} \leq \frac{\pi^2}{6}.$$

This establishes the desired result.

73. (a) $\displaystyle S_{n+1} = S_n + \frac{1}{(n+1)^p} > S_n$ for all $n \geq 1$ since $\displaystyle\frac{1}{(n+1)^p} > 0$ for all $n \geq 1$.

(b) $\displaystyle S_{2m+1} = \sum_{k=1}^{2m+1} \frac{1}{k^p} = 1 + \sum_{k=2}^{2m+1} \frac{1}{k^p} = 1 + \sum_{k=1}^{m} \frac{1}{(2k)^p} + \sum_{k=1}^{m} \frac{1}{(2k+1)^p}$

(c) $\displaystyle S_{2m+1} = 1 + \sum_{k=1}^{m} \frac{1}{(2k)^p} + \sum_{k=1}^{m} \frac{1}{(2k+1)^p} < 1 + \sum_{k=1}^{m} \frac{1}{(2k)^p} + \sum_{k=1}^{m} \frac{1}{(2k)^p} = 1 + 2\sum_{k=1}^{m} \frac{1}{(2k)^p}$

since $\displaystyle\frac{1}{(2k+1)^p} < \frac{1}{(2k)^p}$ for all $k \geq 1$ (since $p > 1$).

(d) $S_{2m+1} < 1 + 2\sum_{k=1}^{m} \dfrac{1}{(2k)^p} = 1 + 2^{1-p}\sum_{k=1}^{m}\dfrac{1}{k^p} = 1 + 2^{1-p}S_m < 1 + 2^{1-p}S_{2m+1}$. (The last inequality holds because the partial sums are strictly increasing.)

(e) By part (d), $S_{2m+1} < 1 + 2^{1-p}S_{2m+1}$. Therefore, $\left(1 - 2^{1-p}\right)S_{2m+1} < 1$ so $S_{2m+1} < \dfrac{1}{1 - 2^{1-p}}$ for every $m \geq 1$. Finally, since any $n \geq 1$ can be written as either $n = 2m$ or $n = 2m + 1$ where $m \geq 1$ is an integer, and since $S_{2m} < S_{2m+1}$ for all $m \geq 1$, it follows that sequence of partial sums is bounded above. Since the sequence of partial sums of the series is increasing and bounded above, this sequence has a finite limit. By definition, this implies that the series converges.

11.3   Testing for Convergence; Estimating Limits

1. (a) When $k \geq 0, k + 2^k \geq 2^k \implies a_k \leq 1/2^k = 2^{-k}$. Since $\displaystyle\sum_{k=0}^{\infty} 2^{-k}$ converges, the comparison test implies

    that $\displaystyle\sum_{k=0}^{\infty} a_k$ converges.

    (b) $R_{10} = \displaystyle\sum_{k=11}^{\infty} a_k < \sum_{k=11}^{\infty} 2^{-k} = 2^{-11} \sum_{k=0}^{\infty} 2^{-k} = 2^{-10}$

    (c) Since $R_{10} < 2^{-10} \approx 0.00097656$, S_n has the desired accuracy if $n \geq 10$.
    $$S_{10} = \sum_{k=0}^{10} a_k = \frac{127807216183}{75344540040} \approx 1.6963.$$

    (d) Since the terms of the series are all positive, the estimate in part (c) *underestimates* the limit.

3. $\displaystyle\sum_{k=2}^{n} a_k < \int_{1}^{n} f(x)\, dx < \sum_{k=1}^{n-1} a_k$ [HINT: Draw pictures like those on pp. 241–243]

5. Draw a picture illustrating the left sum approximation L_n to the integral $\displaystyle\int_{1}^{n+1} a(x)\, dx$. Since the integrand is a decreasing function, R_n overestimates the value of the integral.

7. Draw a picture that illustrates a right Riemann sum approximation to the integral $\displaystyle\int_{n}^{\infty} a(x)\, dx$. The right sum

    is $\displaystyle\sum_{k=n+1}^{\infty} a_k$; it underestimates $\displaystyle\int_{n}^{\infty} a(x)\, dx$ since the integrand is a decreasing function. For the same reason,

    $a_{n+1} + \displaystyle\int_{n+1}^{\infty} a(x)\, dx \leq \int_{n}^{\infty} a(x)\, dx$. Finally, the first inequality established above implies that

    $\displaystyle\sum_{k=n+2}^{\infty} a_k \leq \int_{n+1}^{\infty} a(x)\, dx$. Adding a_{n+1} to both sides of this inequality establishes that

    $\displaystyle\sum_{k+n+1}^{\infty} a_k \leq a_{n+1} + \int_{n+1}^{\infty} a(x)\, dx.$

9. $\displaystyle\int_{1}^{\infty} \frac{dx}{x^{3/2}} = -\frac{2}{\sqrt{x}}\bigg|_{1}^{\infty} = 2 \implies \sum_{k=1}^{\infty} \frac{1}{k\sqrt{k}}$ converges and $2 \leq \displaystyle\sum_{k=1}^{\infty} \frac{1}{k\sqrt{k}} \leq 3.$

11. Since $\displaystyle\int_{2}^{\infty} \frac{dx}{(\ln x)^2} \geq \int_{2}^{\infty} \frac{dx}{4x} = \infty$, the series diverges by the integral test.

13. (a) The integral test can't be used to prove that the series converges because the function $a(x) = \dfrac{2 + \sin x}{x^2}$ is not decreasing on the interval $[1, \infty)$.

    (b) Since $\dfrac{2 + \sin k}{k^2} \leq \dfrac{3}{k^2}$ for all $k \geq 1$, and $\displaystyle\sum_{k=1}^{\infty} \frac{3}{k^2}$ converges, the comparison test implies that the series

    $\displaystyle\sum_{k=1}^{\infty} \frac{2 + \sin k}{k^2}$ converges.

15. $\displaystyle\lim_{k \to \infty} \frac{\frac{2^{k+1}}{(k+1)!}}{\frac{2^k}{k!}} = \lim_{k \to \infty} \frac{2}{k+1} = 0 \implies \sum_{k=1}^{\infty} \frac{2^k}{k!}$ converges.

17. $\lim\limits_{m \to \infty} \dfrac{\frac{(m+1)!}{(2m+2)!}}{\frac{m!}{(2m)!}} = \lim\limits_{m \to \infty} \dfrac{m+1}{(2m+2)(2m+1)} = 0$ so $\sum\limits_{m=1}^{\infty} \dfrac{m!}{(2m)!}$ converges.

19. (a) $S_{n+1} = \sum\limits_{k=1}^{n+1} a_k = S_n + a_k \geq S_n$ since $a_k = a(k) \geq 0$ for all integers $k \geq 1$.

   (b) $\int_1^n a(x)\,dx \leq \int_1^{\infty} a(x)\,dx$ because $a(x) \geq 0$ for all $x \geq 1$.

   (c) Part (a) implies that $\{S_n\}$ is an increasing sequence. Since $S_n \leq \int_1^n a(x)\,dx$, part (b) implies that the sequence of partial sums is bounded above. Thus, the sequence of partial sums converges to a limit.

21. Yes. Let the series in question be $\sum\limits_{k=0}^{\infty} a_k$ (i.e., $a_0 = 1$, $a_1 = 1/3$, $a_2 = 1/15$, $a_3 = 1/85$, etc.). Then

   $a_k \leq 1/(2k-1)(2k+1) = 1/(4k^2-1) \leq 1/3k^2$ for all $k \geq 1$. Since $\sum\limits_{k=1}^{\infty} \dfrac{1}{3k^2}$ converges, $\sum\limits_{k=0}^{\infty} a_k$ also converges

   (by the comparison test).

   NOTE: The given series is $\sum\limits_{k=0}^{\infty} a_k$ with $a_k = \dfrac{2^k k!}{(2k+1)!}$. Therefore, the ratio test can also be used to show that the series converges.

23. (a) H_n is the nth partial sum of the harmonic series. These partial sums form a monotonically increasing, divergent sequence.

   (b) $S_n = \sum\limits_{k=0}^{n} \dfrac{1}{2k+1} \geq \sum\limits_{k=0}^{n} \dfrac{1}{2k+2} = \dfrac{1}{2}\sum\limits_{k=0}^{n} \dfrac{1}{k+1} = \dfrac{1}{2}\sum\limits_{k=1}^{n-1} \dfrac{1}{k} = \dfrac{1}{2}H_{n+1} > \dfrac{1}{2}H_n$

   (c) $\sum\limits_{k=0}^{\infty} \dfrac{1}{2k+1}$ diverges.

25. The improper integral $\int_2^{\infty} \dfrac{dx}{x(\ln n)^p}$ converges when $p > 1$ and diverges when $p \leq 1$. Thus, by the integral test, the series converges only for $p > 1$.

27. (a) $\lim\limits_{n \to \infty} \dfrac{a_{n+1}}{a_n}$ does not exist. For $m = 1, 2, 3, \ldots$, $\dfrac{a_{2m}}{a_{2m-1}} = \left(\dfrac{2}{3}\right)^m$ and $\dfrac{a_{2m+1}}{a_{2m}} = \dfrac{1}{2}\left(\dfrac{3}{2}\right)^m$.

   (b) Because the limit in part (a) does not exist, the ratio test says nothing about the convergence of the series

   $\sum\limits_{k=1}^{\infty} a_k$.

   (c) $\sum\limits_{k=1}^{\infty} a_k = \sum\limits_{k=1}^{\infty} 2^{-k} + \sum\limits_{k=1}^{\infty} 3^{-k} = 1 + \dfrac{1}{2} = \dfrac{3}{2}$

29. The harmonic series $\sum\limits_{k=1}^{\infty} \dfrac{1}{k}$ has the specified properties.

31. Let $a_n = \dfrac{n^n}{n!}$. Then $\dfrac{a_{n+1}}{a_n} = \dfrac{(n+1)^{n+1}}{n^n} \dfrac{n!}{(n+1)!} = \left(\dfrac{n+1}{n}\right)^n$ and $\lim\limits_{n \to \infty} \dfrac{a_{n+1}}{a_n} = e > 1$. Therefore, the series

   $\sum\limits_{n=1}^{\infty} \dfrac{n^n}{n!}$ diverges.

33. $\displaystyle\sum_{j=0}^{\infty} \frac{1}{j+e^j} < \sum_{j=0}^{\infty} \frac{1}{e^j} = \frac{1}{1-e^{-1}}.$ (The series on the right is a convergent geometric series.)

35. $\displaystyle\sum_{k=1}^{\infty} \frac{k}{\left(k^2+1\right)^2} < \sum_{k=1}^{\infty} \frac{1}{k^3} \leq 1 + \int_1^{\infty} x^{-3}\, dx = \frac{3}{2}.$

37. Converges—comparison test: $\displaystyle\sum_{n=1}^{\infty} \frac{\arctan n}{1+n^2} < \frac{\pi}{2} \sum_{n=1}^{\infty} \frac{1}{1+n^2} < \frac{\pi}{2} \sum_{n=1}^{\infty} \frac{1}{n^2} = \frac{\pi^3}{12}.$

   [NOTE: This series can also be shown to converge via the integral test: $\int_1^{\infty} \frac{\arctan x}{1+x^2}\, dx = \int_{\pi/4}^{\pi/2} u\, du = \frac{3\pi^2}{32}$. This

   leads to the bound $\displaystyle\sum_{n=1}^{\infty} \frac{\arctan n}{1+n^2} \leq \frac{\pi}{8} + \frac{3\pi^2}{32}.$]

39. Diverges—integral test: $\displaystyle\int_1^{\infty} \frac{dx}{100+5x} = \frac{1}{5} \int_{105}^{\infty} \frac{du}{u} = \infty.$

41. Converges—comparison test: $\displaystyle\sum_{n=1}^{\infty} \frac{1}{n\, 3^n} < \sum_{n=1}^{\infty} \frac{1}{3^n} = \frac{1}{2}.$

43. Converges—integral test: $\displaystyle\sum_{j=1}^{\infty} j5^{-j} \leq \frac{1}{5} + \int_1^{\infty} x5^{-x}\, dx = \frac{1}{5} + \frac{1+\ln 5}{5(\ln 5)^2}.$

   NOTE: This series can also be shown to converge via the ratio test:

$$\lim_{j\to\infty} \frac{\frac{j+1}{5^{j+1}}}{\frac{j}{5^j}} = \lim_{j\to\infty} \frac{j+1}{j} \cdot \frac{5^j}{5^{j+1}} = \lim_{j\to\infty} \frac{j+1}{5j} = \frac{1}{5} < 1.$$

   Furthermore, since $a_{j+1}/a_j = (j+1)/5j \leq 2/5$ when $j \geq 1$, $\displaystyle\sum_{j=1}^{\infty} j5^{-j} \leq \sum_{j=1}^{\infty} \frac{2^{j-1}}{5^j} = \frac{1}{3}.$

45. Diverges. The given series is 2 minus the harmonic series: $1 - \dfrac{1}{2} - \dfrac{1}{3} - \dfrac{1}{4} - \dfrac{1}{5} - \cdots = 2 - \displaystyle\sum_{k=1}^{\infty} \frac{1}{k}$. Since the

   harmonic series diverges, so does this series.

47. Diverges—comparison test: $\displaystyle\sum_{n=2}^{\infty} \frac{1}{\sqrt[3]{n^2-1}} > \sum_{n=2}^{\infty} \frac{1}{n^{2/3}}.$

49. Converges—integral test: $\displaystyle\int_0^{\infty} e^{-x^2}\, dx = \sqrt{\pi}/2. \ \sum_{m=0}^{\infty} e^{-m^2} \leq 1 + \int_0^{\infty} e^{-x^2}\, dx = 1 + \sqrt{\pi}/2.$

51. Converges—ratio test: $\displaystyle\lim_{n\to\infty} \frac{\frac{(n+1)!}{(2n+2)!}}{\frac{n!}{(2n)!}} = \lim_{n\to\infty} \frac{(n+1)!}{n!} \cdot \frac{(2n)!}{(2n+2)!} = \lim_{n\to\infty} \frac{1}{2(2n+1)} = 0.$

   Since $\dfrac{a_{n+1}}{a_n} = \dfrac{1}{2(2n+1)} \leq \dfrac{1}{2}$ for all $n \geq 0$, $\displaystyle\sum_{n=0}^{\infty} a_n \leq a_0 \sum_{n=0}^{\infty} \frac{1}{2^n} = 2a_0 = 2.$

   NOTE: Since $\dfrac{n!}{(2n)!} \leq \dfrac{1}{(n+1)!}$ when $n \geq 0$, the series can also be shown to converge using the comparison

   test: $\displaystyle\sum_{n=0}^{\infty} \frac{n!}{(2n)!} < \sum_{n=0}^{\infty} \frac{1}{(n+1)!} = e - 1.$

53. Diverges—comparison test: $\displaystyle\sum_{k=1}^{\infty} \frac{k!}{(k+1)!-1} > \sum_{k=1}^{\infty} \frac{k!}{(k+1)!} = \sum_{k=1}^{\infty} \frac{1}{k+1} = \frac{1}{2} + \frac{1}{3} + \frac{1}{4} + \cdots$

55. Diverges by the nth term test: $\displaystyle\lim_{n\to\infty}\sum_{k=1}^{n}k^{-1}=\infty$

57. Converges (by the comparison test): $\displaystyle\sum_{k=1}^{\infty}\frac{1}{k^2+3}\le\sum_{k=1}^{\infty}\frac{1}{k^2}$. (The series on the right side of the inequality is a convergent p-series.)

Since $\displaystyle R_N=\sum_{k=N+1}^{\infty}\frac{1}{k^2+3}\le\sum_{k=N+1}^{\infty}\frac{1}{k^2}\le\int_N^\infty\frac{dx}{x^2}=\frac{1}{N}\le0.001$ when $N\ge1000$, $n\ge1000$ implies that S_n approximates the sum of the series within 0.001.

59. Converges: $\displaystyle\sum_{j=2}^{\infty}\frac{3^j}{4^{j+1}}=\frac{1}{4}\sum_{j=2}^{\infty}\left(\frac{3}{4}\right)^j=\frac{9}{64}\sum_{j=0}^{\infty}\left(\frac{3}{4}\right)^j=\frac{9}{64}\frac{1}{1-3/4}=\frac{9}{16}$.

$\displaystyle R_N=\sum_{j=N+1}^{\infty}\frac{3^j}{4^{j+1}}=\frac{3^{N+1}}{4^{N+2}}\sum_{j=0}^{\infty}\left(\frac{3}{4}\right)^j=\left(\frac{3}{4}\right)^{N+1}\le0.001$ when $N\ge-\dfrac{\ln1000}{\ln(3/4)}-1\approx23.012$. Thus, $n\ge24$ implies that S_n approximates the sum of the series within 0.001.

61. Converges.

$0<R_n=\displaystyle\sum_{m=n+1}^{\infty}\frac{\ln m}{m^3}\le\int_n^\infty\frac{\ln x}{x^3}dx\le\int_n^\infty\frac{x}{x^3}dx=\frac{1}{n}\le0.001$ when $n\ge1000$.

$0<R_n=\displaystyle\sum_{m=n+1}^{\infty}\frac{\ln m}{m^3}\le\int_n^\infty\frac{\ln x}{x^3}dx\le\int_n^\infty\frac{\sqrt{x}}{x^3}dx=\frac{2}{3n^{3/2}}\le0.001$ when $n\ge77$.

$0<R_n=\displaystyle\sum_{m=n+1}^{\infty}\frac{\ln m}{m^3}\le\int_n^\infty\frac{\ln x}{x^3}dx=\frac{1+2\ln n}{4n^2}\le0.001$ when $n\ge47$.

63. Since the improper integral $\displaystyle\int_2^\infty\frac{1}{x(\ln x)^5}dx=\int_{\ln 2}^\infty\frac{1}{u^5}=\frac{1}{4(\ln2)^4}$ converges, the corresponding series converges. Furthermore, since $\displaystyle\int_n^\infty\frac{dx}{x(\ln x)^5}dx<0.005$ when $n\ge15$, the approximation $L\approx\displaystyle\sum_{k=2}^{15}\frac{1}{k(\ln k)^5}\approx3.4254$ has the desired accuracy.

65. (a) $k!=\overbrace{k\cdot(k-1)\cdots\cdots2}^{(k-1)\text{-terms}}\ge\overbrace{2\cdot2\cdots\cdots2}^{(k-1)\text{-terms}}=2^{k-1}\implies\dfrac{1}{k!}\le\dfrac{1}{2^{k-1}}$

 (b) $k!=\overbrace{k\cdot(k-1)\cdots\cdots(k-10)\cdot10\cdots\cdots2}^{(k-10)\text{-terms}}\ge\overbrace{10\cdot10\cdots\cdots10\cdot10!}^{(k-10)\text{-terms}}=10^{k-1}\cdot10!\implies\dfrac{1}{k!}\le\dfrac{1}{10!\,10^{k-10}}$

 (c) Let $S_n=\displaystyle\sum_{k=0}^{n}\frac{1}{k!}$. S_{10} underestimates $\displaystyle\sum_{k=0}^{\infty}\frac{1}{k!}$ since S_n is a monotonically increasing sequence.

$R_{10}=\displaystyle\sum_{k=0}^{\infty}\frac{1}{k!}-S_{10}=\sum_{k=11}^{\infty}\frac{1}{k!}\le\sum_{k=11}^{\infty}\frac{1}{10!\,10^{k-10}}=\frac{1}{10!}\sum_{k=1}^{\infty}\frac{1}{10^k}=\frac{1}{10!}\frac{1}{9}=\frac{1}{32659200}\approx3.0619\times10^{-8}$

67. (a) Let $a_k=(1/k!)^2$. Since $a_{k+1}/a_k=\left(k!/(k+1)!\right)^2=1/(k+1)^2$, $\displaystyle\lim_{k\to\infty}\frac{a_{k+1}}{a_k}=0$. Thus, the ratio test implies that $\displaystyle\sum_{k=1}^{\infty}a_k$ converges.

(b) Let $R_N = \displaystyle\sum_{k=N+1}^{\infty} \left(\frac{1}{k!}\right)^2$. Now,

$$R_N = \sum_{k=0}^{\infty} \left(\frac{1}{(N+1+k)!}\right)^2$$

$$< \sum_{k=0}^{\infty} \left(\frac{1}{(N+1)^k (N+1)!}\right)^2 = \frac{1}{((N+1)!)^2} \sum_{k=0}^{\infty} \left(\frac{1}{(N+1)^2}\right)^k$$

$$= \frac{1}{((N+1)!)^2} \frac{1}{1-\left(\frac{1}{N+1}\right)^2}.$$

Therefore, since $R_N < 5 \times 10^{-6}$ when $N \geq 5$, S_N approximates the sum of this series within 5×10^{-6} when $N \geq 5$.

69. (a) If $k \geq N$, then $k! = \overbrace{k \cdot (k-1) \cdot (k-2) \cdots \cdot (N+1)}^{(k-N)\text{-terms}} \cdot N! \geq (N+1)^{k-N} \cdot N!$.

Thus, $\dfrac{x^k}{k!} = \dfrac{x^N \cdot x^{k-N}}{k!} \leq \dfrac{x^N \cdot x^{k-N}}{N! \cdot (N+1)^{k-N}} = \dfrac{x^N}{N!}\left(\dfrac{x}{N+1}\right)^{k-N}$.

(b) $\displaystyle\sum_{k=N}^{\infty} \frac{x^k}{k!} \leq \sum_{k=N}^{\infty} \frac{x^N}{N!}\left(\frac{x}{N+1}\right)^{k-N} = \frac{x^N}{n!}\sum_{k=N}^{\infty}\left(\frac{x}{N+1}\right)^{k-N} = \frac{x^N}{n!}\sum_{k=0}^{\infty}\left(\frac{x}{N+1}\right)^k = \frac{x^N}{N!} \cdot \frac{1}{1-\frac{x}{N+1}}$

since $x/(N+1) < 1$.

71. There does not exist a number $r < 1$ such that $\dfrac{a_{k+1}}{a_k} \leq r$ for all $k \geq 1$.

73. Let $a_n = (n!)^2/(2n)!$. Since $a_{n+1}/a_n = \frac{n+1}{2(2n+1)} \leq \frac{1}{3}$ when $n \geq 1$, $R_N = \displaystyle\sum_{n=N+1}^{\infty} a_n \leq a_{N+1}\sum_{n=0}^{\infty}\left(\frac{1}{3}\right)^n = \frac{3}{2}a_{N+1}$.

Thus, $R_N < 0.0005$ when $N \geq 6$.

75. (a) $\displaystyle\int_x^{\infty} f'(t)\,dt = \lim_{a\to\infty}\int_x^a f'(t)\,dt = \lim_{a\to\infty}\left(f(a)-f(x)\right) = -f(x)$.

[NOTE: $f(a) = \ln\left(\dfrac{a+1}{a}\right) - \dfrac{1}{a+1} = \ln\left(1+\dfrac{1}{a}\right) - \dfrac{1}{a+1}$.]

(b) When $x > 0$, $f(x) = -\displaystyle\int_x^{\infty} f'(t)\,dt > \int_x^{\infty} \frac{dt}{(t+1)^3} = \frac{1}{2(x+1)^2}$.

(c) Let $S_N = \displaystyle\sum_{k=n}^{N}(a_k - a_{k+1}) = (a_n - a_{n+1}) + (a_{n+1} - a_{n+2}) + \cdots + (a_N - a_{N+1}) = a_n - a_{N+1}$.

Since $\gamma = \displaystyle\lim_{n\to\infty} a_n$, $\displaystyle\sum_{k=n}^{\infty}(a_k - a_{k+1}) = \lim_{N\to\infty} S_N = a_n - \lim_{N\to\infty} a_{N+1} = a_n - \gamma$.

(d) Since f is a decreasing function, the integral test implies that $\displaystyle\int_n^{\infty} f(x)\,dx \leq \sum_{k=n}^{\infty} f(k)$. Therefore, part (b)

implies that $\displaystyle\int_n^{\infty} f(x)\,dx > \frac{1}{2(n+1)}$.

To get the upper bound on $a_n - \gamma$, note that $f(k) < \dfrac{1}{2}\left(\dfrac{1}{k} - \dfrac{1}{k+1}\right)$. (Apply the trapezoid rule to

$\int_k^{k+1} dx/x$.) This inequality implies that $\displaystyle\sum_{k=n}^{\infty} f(k) < \sum_{k=n}^{\infty}\frac{1}{2}\left(\frac{1}{k}-\frac{1}{k+1}\right) = \frac{1}{2n}$.

NOTE: An argument similar to that used in part (b) shows that $f(x) < 1/2x^2$. Therefore, the integral test implies that $\displaystyle\sum_{k=n}^{\infty} f(k) \le \int_{n-1}^{\infty} f(x)\,dx < \int_{n-1}^{\infty} \frac{1}{2x^2}\,dx = \frac{1}{2(n-1)}$ which is not quite the desired result.

The lower bound on $a_n - \gamma$ can also be derived in the following way: Applying the midpoint rule to $\int_k^{k+1} dx/x$ yields $f(n) > \dfrac{1}{n+\frac{1}{2}} - \dfrac{1}{n+1} = \dfrac{1}{2n^2+3n+1} > \dfrac{1}{2(n^2+3n+2)} = \dfrac{1}{2}\left(\dfrac{1}{n+1} - \dfrac{1}{n+2}\right)$.

Therefore, $\displaystyle\sum_{k=n}^{\infty} f(k) > \sum_{k=n}^{\infty} \frac{1}{2}\left(\frac{1}{k+1} - \frac{1}{k+2}\right) = \frac{1}{2(n+1)}$.

77. Let $a_k = 1/k$. Then $\displaystyle\sum_{k=1}^{\infty} 2^k a_{2^k} = \sum_{k=1}^{\infty} 1$ which is obviously a divergent series. Therefore, part (c) of the previous exercise implies that $\displaystyle\sum_{k=1}^{\infty} a_k$ diverges.

11.4   Absolute Convergence; Alternating Series

1. (a) The series converges conditionally. (After the fifth term, the series has the same terms as the alternating harmonic series.)

   (b) $S_{15} = 1 + 2 + 3 + 4 + 5 + \displaystyle\sum_{k=6}^{15} \frac{(-1)^{k+1}}{k} = 15 - \frac{20887}{360360} \approx 14.942$. S_{15} *overestimates* S because the last term included in the alternating series was positive.

   (c) $14.902 < S < 14.902 + \frac{1}{61} \approx 14.918$

   (d) $S = 15 + \left(\ln 2 - \displaystyle\sum_{k=1}^{5} \frac{(-1)^{k+1}}{k} \right) = 15 + \ln 2 - \frac{47}{60} \approx 14.910$

3. (a) $0 < \displaystyle\sum_{k=1}^{\infty} \frac{|a_k|}{k} < \sum_{k=1}^{\infty} |a_k|$ so the series $\displaystyle\sum_{k=1}^{\infty} \frac{a_k}{k}$ converges absolutely. This implies (by Theorem 9) that this series converges.

   (b) No — An example is $a_k = 1/k$. Then, $\displaystyle\sum_{k=1}^{\infty} a_k/k = \sum_{k=1}^{\infty} 1/k^2$ (a convergent series), but $\displaystyle\sum_{k=1}^{\infty} a_k = \sum_{k=1}^{\infty} 1/k$ (the harmonic series).

5. The series converges absolutely by the alternating series test. Since $c_{n+1} = (n+1)^{-4} < 0.005$ when $n = 3$, $\left| S - \displaystyle\sum_{k=1}^{N} (-1)^k/k^4 \right| < 0.005$ when $N \geq 3$. Using $N = 3$, $S \approx -\frac{1231}{1296} \approx -0.94985$.

7. The series converges absolutely by the comparison test using $b_k = (2/7)^k$. $\left| S - \displaystyle\sum_{k=0}^{N} a_k \right| \leq 0.005$ when $N \geq 4$ since $2^5/(7^5 + 5) < 0.005$. Using $N = 4$, $S \approx \frac{68917177}{84877260} \approx 0.81196$.

9. The series converges absolutely by the ratio test. $\left| S - \displaystyle\sum_{k=5}^{N} a_k \right| \leq 0.005$ when $N \geq 13$ since $14^{10}/10^{14} < 0.005$. Using $N = 13$, $S \approx -\frac{573982077919709}{10000000000000} \approx -57.398$.

11. No. Because $\displaystyle\lim_{n \to \infty} (-1)^n \frac{n}{2n - 1}$ does not exist, the series cannot converge (by the n-th term test).

13. (a) When $p \leq 0$, $\displaystyle\lim_{k \to \infty} \frac{\ln k}{k^p} = \infty$, so the series diverges by the nth term test (Theorem 5, p. 230).

   When $p > 0$, the function $f(x) = \dfrac{\ln x}{x^p}$ is continuous, positive, and decreasing on $(e^{1/p}, \infty)$ Therefore, the integral test (Theorem 7, p. 243) implies that the series converges if and only if the improper integral $\displaystyle\int_{e^{1/p}}^{\infty} f(x)\, dx$ converges.

   When $p \neq 1$, $\displaystyle\int \frac{\ln x}{x^p}\, dx = \frac{(1 - p)\ln x - 1}{(1 - p)^2 x^{p-1}}$ and, $\displaystyle\int \frac{\ln x}{x}\, dx = \frac{1}{2}(\ln x)^2$. Therefore, $\displaystyle\int_{e^{1/p}}^{\infty} \frac{\ln x}{x^p}\, dx$ diverges when $p \leq 1$ and converges when $p > 1$. It follows that the series converges only when $p > 1$.

   (b) The alternating series test (Theorem 10, p. 255) implies that the series converges for every $p > 0$. (When $p > 0$, the terms of the series decrease in magnitude for $k > e^{1/p}$, approach zero, and alternate in sign.)

   (c) The series converges absolutely when $p > 1$ since $\displaystyle\sum_{k=2}^{\infty} \frac{\ln k}{k^p}$ converges only when $p > 1$.

(d) When $0 < p \leq 1$, $\displaystyle\sum_{k=2}^{\infty} \frac{\ln k}{k^p}$ diverges but $\displaystyle\sum_{k=2}^{\infty}(-1)^k \frac{\ln k}{k^p}$ converges. Therefore, the series converges conditionally when $0 < p \leq 1$.

15. converges absolutely—$\displaystyle\sum_{j=1}^{\infty} \frac{1}{j^2}$ is a convergent p-series $(p = 2)$. $\dfrac{3}{4} < \displaystyle\sum_{j=1}^{\infty} \frac{(-1)^{j+1}}{j^2} < 1$

17. converges conditionally—$\displaystyle\sum_{k=4}^{\infty} \frac{\ln k}{k}$ diverges by the integral test but the terms of the series form a decreasing

sequence and $\displaystyle\lim_{k\to\infty} \frac{\ln k}{k} = 0$. $\dfrac{\ln 4}{4} - \dfrac{\ln 5}{5} < \displaystyle\sum_{k=4}^{\infty}(-1)^k \frac{\ln k}{k} < \dfrac{\ln 4}{4}$

19. converges conditionally—$\displaystyle\sum_{n=1}^{\infty} \frac{\cos(n\pi)}{n} = \sum_{n=1}^{\infty} \frac{(-1)^n}{n}$ which is (almost) the alternating harmonic series.

$-1 < \displaystyle\sum_{n=1}^{\infty} \frac{\cos(n\pi)}{n} = -\ln 2 < -1/2$

21. converges absolutely—Let $a_m = 4m^3/2^m$. Then $\displaystyle\lim_{m\to\infty} \frac{a_{m+1}}{a_m} = \lim_{m\to\infty} \frac{(m+1)^3}{2m^3} = \frac{1}{2} < 1$ so $\displaystyle\sum_{m=0}^{\infty} \frac{4m^3}{2^m}$ con-

verges by the ratio test. The terms of the series are decreasing in absolute value for all $m \geq 4$. Thus,

$\displaystyle\sum_{m=0}^{5}(-1)^m a_m = -\frac{57}{8} < \sum_{m=0}^{\infty}(-1)^m \frac{4m^3}{2^m} < \sum_{m=0}^{4}(-1)^m a_m = \frac{17}{2}$

23. converges absolutely—$\displaystyle\lim_{j\to\infty} \frac{a_{j+1}}{a_j} = \lim_{j\to\infty} \frac{j+1}{(j^2+2j+1)\cdots(j^2+1)} = 0$. The terms of the series are decreas-

ing in absolute value for all $j \geq 1$. Thus, $\displaystyle\sum_{j=0}^{1}(-1)^j a_j = 0 < \sum_{j=0}^{\infty}(-1)^j a_j < \sum_{j=0}^{2}(-1)^j a_j = \frac{1}{12}$.

25. Let $b_k = a_{k+10^9}$. The alternating series test can be used to show that $\displaystyle\sum_{k=1}^{\infty}(-1)^{k+1} b_k$ converges. Since

$\displaystyle\sum_{k=1}^{\infty}(-1)^{k+1} a_k = \sum_{k=1}^{10^9}(-1)^{k+1} a_k + \sum_{k=10^9+1}^{\infty}(-1)^{k+1} a_k = \sum_{k=1}^{10^9}(-1)^{k+1} a_k + \sum_{k=1}^{\infty}(-1)^{k+1} b_k$, the series $\displaystyle\sum_{k=1}^{\infty}(-1)^{k+1} a_k$
converges.

27. (a) Theorem 9 (p. 253) implies that $\displaystyle\sum_{j=1}^{\infty}(-1)^{j+1} b_j$ converges absolutely.

(b) Since the terms of the series defining S are all positive, $S - \sum_{j=1}^{100} b_j \leq 0.005$ implies that $b_j \leq 0.005$ for all $j \geq 101$. Therefore, since $b_{j+1} \leq b_j$, Theorem 10 implies the desired result.
Alternatively,

$$\left| \sum_{j=1}^{\infty}(-1)^{j+1} b_j - \sum_{j=1}^{100}(-1)^{j+1} b_j \right| = \left| \sum_{j=101}^{\infty}(-1)^{j+1} b_j \right| \leq \sum_{j=101}^{\infty} b_j \leq 0.005.$$

29. No. Since $a_k \geq 0$ for all $k \geq 1$, $|a_k| = a_k$ for all $k \geq 1$.

31. (a) Let $a_k = (-1)^{k+1}/k^2$ and $b_k = 1/k^2$. Then $a_{2m-1} = b_{2m-1}$ and $a_{2m} < b_{2m}$ for $m = 1, 2, 3, \ldots$. Thus,

$\displaystyle\sum_{k=n+1}^{\infty} a_k \leq \sum_{k=n+1}^{\infty} b_k$.

(b) According to part (a), for each n the tail of the alternating series is smaller than the tail of the series of positive terms. Thus, the error made by approximating the alternating series by its nth partial sum is less than the error made by approximating the series with positive terms by its nth partial sum.

33. (b) Because of the $\cot k$ term, there is not a value of N such that the ratio is less than 1 for all $k > N$.

(c) The series converges absolutely since $|a_k| \leq 1/k^2$ and $\sum_{k=1}^{\infty} \frac{1}{k^2}$ converges.

11.5   Power Series

1. $P_1(x) = x$, $P_2(x) = P_1(x) + x^2/2$, $P_4(x) = P_2(x) + +x^3/3 + x^4/4$, $P_6(x) = P_4(x) + x^5/5 + x^6/6$, $P_8(x) = P_6(x) + x^7/7 + x^8/8$, $P_{10}(x) = P_8(x) + x^9/9 + x^{10}/10$.

3. $\left|\dfrac{a_{k+1}}{a_k}\right| = \dfrac{k}{2(k+1)} \cdot |x| \implies R = 2$

5. $\left|\dfrac{a_{m+1}}{a_m}\right| = \dfrac{m^2+1}{(m+1)^2+1} \cdot |x| \implies R = 1$

7. $\left|\dfrac{x^n}{n!+n}\right| \le \dfrac{|x|^n}{n!} \implies R = \infty$

9. $R = \infty$; interval of convergence is $(-\infty, \infty)$

11. $R = 1/3$; interval of convergence is $[-1/3, 1/3]$

13. $R = 1$; interval of convergence is $[2, 4]$

15. $R = 1$; interval of convergence is $[-2, 0)$

17. (a) $\displaystyle\sum_{k=1}^{\infty} \frac{x^k}{k4^k}$

　　(b) $\displaystyle\sum_{k=1}^{\infty} \frac{(x-2)^k}{k2^{3k}}$

　　(c) $\displaystyle\sum_{k=1}^{\infty} \frac{(x+2)^k}{2^k}$

　　(d) $\displaystyle\sum_{k=1}^{\infty} \frac{(12-x)^k}{k4^k}$

　　(e) $\displaystyle\sum_{k=1}^{\infty} \frac{(x+7)^k}{k4^k}$

19. (a) By definition, the radius of convergence of a power series is the largest value of R such that the series converges for all x such that $|x| < R$.

　　(b) This power series converges when $|x - 1| < 2$ and diverges when $|x - 1| > 2$. Thus, its radius of convergence $R = 2$.

　　(c) Let $z = x - 3$. Since $\displaystyle\sum_{k=0}^{\infty} z^k$ converges only when $-2 < z \le 2$, $\displaystyle\sum_{k=0}^{\infty} a_k(x-3)^k$ converges only when $1 < x \le 5$.

　　(d) The power series $\displaystyle\sum_{k=0}^{\infty} a_k(x+1)^k$ converges only when $-2 < x+1 \le 2$. Thus, its interval of convergence is $(-3, 1]$.

21. $[1, 5)$

23. $[-3, 5)$

25. $[-6, -4]$

27. The information given implies that the power series converges on the interval $[-3, 3)$, diverges when $x \ge 7$, and diverges when $x < -7$. It does not imply anything about convergence or divergence on the intervals $[-7, -3)$ and $[3, 7)$.

(a) cannot

(b) may

(c) may

(d) cannot

(e) may

(f) may

29. (a) Cannot be true. The interval of convergence of a power series is symmetric around and includes its base point ($b = 1$ in this case).

(b) May be true. (The statement is true when $a_k = 1/k!$ but it is false when $a_k = 1$.)

(c) Must be true. If the radius of convergence of the power series is 3, then the interval of convergence includes all values of x such that $|x - 1| < 3$.

(d) Cannot be true. The interval of convergence of this power series must be symmetric about the point $b = 1$.

(e) Cannot be true. The interval of convergence of the power series is the solution set of the inequality $|x - 1| < 8$. Thus, the radius of convergence of the power series is 8.

31. (a) The series $\sum_{n=0}^{\infty} \dfrac{2 \cdot 10^n}{3^n + 5}$ diverges by the ratio test—the limit of the ratio of successive terms of the series is $10/3 > 1$.

(b) The power series defining f converges when $-3 < x < 3$. Thus, only 0.5 and 1.5 are in the domain of f.

(c) $f(1) - \sum_{n=0}^{N} \dfrac{2}{3^n + 5} = \sum_{n=N+1}^{\infty} \dfrac{2}{3^n + 5} < \sum_{n=N+1}^{\infty} \dfrac{2}{3^n} = \dfrac{1}{3^{N+1}} \sum_{n=0}^{\infty} \dfrac{2}{3^n} = \dfrac{1}{3^{N+1}} \cdot 3 = \dfrac{1}{3^N}$. Thus, since $3^{-5} <$ 0.01, $\sum_{n=0}^{5} \dfrac{2}{3^n + 5} = \dfrac{367273}{447888} \approx 0.82001$ approximates $f(1)$ within 0.01.

33. (a) The domain of g is $[-9, 1]$.

(b) $g(0) - \sum_{n=1}^{N} \dfrac{4^n}{n^3 5^n} = \sum_{n=N+1}^{\infty} \dfrac{4^n}{n^3 5^n} < \int_{N}^{\infty} \dfrac{dx}{x^3} = \dfrac{1}{2N^2} \leq 0.005$ when $N \geq 10$. Thus,

$$g(0) \approx \sum_{n=0}^{10} \dfrac{4^n}{n^3 5^n} = \dfrac{277892997449134}{305233154296875} \approx 0.91043.$$

(c) The approximation $g(-5) \approx -\dfrac{1}{5}$ is correct within 0.005 because the series defining $g(-5)$ is an alternating series. Since the magnitude of the second term in the series is 0.005, the error made by approximating the series by its first term is smaller than 0.005.

11.6 Power Series as Functions

1. (a) Since $\left|\dfrac{a_{k+1}}{a_k}\right| = \dfrac{|x|}{2}$, the radius of convergence is 2.

   (b) Since $\left|\dfrac{a_{k+1}}{a_k}\right| = \dfrac{(k+1)|x|}{2k}$, the radius of convergence is 2.

   (c) Since $\left|\dfrac{a_{k+1}}{a_k}\right| = \dfrac{(k+1)|x|}{2(k+2)}$, the radius of convergence is 2.

3. $f(x) = \left(1 - x^2\right)^{-1} = \displaystyle\sum_{k=0}^{\infty} x^{2k}$ [Substitute $u = x^2$ into the power series representation of $(1-u)^{-1}$.]

5. $f(x) = \dfrac{x}{1 - x^4} = x \displaystyle\sum_{k=0}^{\infty} x^{4k} = \sum_{k=0}^{\infty} x^{4k+1}$

7. $\cos(x^2) = \displaystyle\sum_{k=0}^{\infty} (-1)^k \dfrac{x^{4k}}{(2k)!}$

9. $\ln\left(1 + \sqrt[3]{x}\right) = \displaystyle\sum_{k=1}^{\infty} (-1)^{k+1} \dfrac{x^{k/3}}{k}$

11. $1/\sqrt{e} = e^{-1/2} = \displaystyle\sum_{k=0}^{\infty} \dfrac{(-1/2)^k}{k!}$. Since $1/(2^4 \cdot 4!) < 0.005$, $\displaystyle\sum_{k=0}^{3} \dfrac{(-1/2)^k}{k!} = \dfrac{29}{48} \approx 0.60417 \approx 1/\sqrt{e}$ within 0.005.

13. Let $f(x) = 1/(1-x) = \displaystyle\sum_{n=0}^{\infty} x^n$ if $|x| < 1$. Then $f'(x) = 1/(1-x)^2 = \displaystyle\sum_{n=1}^{\infty} n x^{n-1} = \dfrac{1}{x}\sum_{n=1}^{\infty} n x^n$ if $|x| < 1$ and $x \neq 0$. Therefore, $\displaystyle\sum_{n=1}^{\infty} \dfrac{n}{2^n} = (1/2) f'(1/2) = 2$.

15. $x - \sin x = \displaystyle\sum_{k=1}^{\infty} (-1)^{k+1} \dfrac{x^{2k+1}}{(2k+1)!} = \dfrac{x^3}{3!} - \dfrac{x^5}{5!} \pm \cdots = x^3\left(\dfrac{1}{3!} - \dfrac{x^2}{5!} \pm \cdots\right)$

   $(x \sin x)^{3/2} = \left(\displaystyle\sum_{k=0}^{\infty} (-1)^k \dfrac{x^{2k+2}}{(2k+1)!}\right)^{3/2} = \left(x^2 - \dfrac{x^4}{3!} \pm \cdots\right)^{3/2}$

   $= \left(x^6 - \dfrac{1}{2}x^8 \pm \cdots\right)^{1/2} = x^3\left(1 - \dfrac{1}{2}x^2 \pm \cdots\right)^{1/2}$

   Thus,

$$\lim_{x \to 0^+} \frac{x - \sin x}{(x \sin x)^{3/2}} = \lim_{x \to 0^+} \frac{x^3\left(\frac{1}{3!} - \frac{x^2}{5!} \pm \cdots\right)}{x^3\left(1 - \frac{1}{2}x^2 \pm \cdots\right)^{1/2}} = \frac{1}{6}.$$

17. $\dfrac{e^x - 1}{x} = x^{-1}\left(\displaystyle\sum_{k=0}^{\infty} \dfrac{x^k}{k!} - 1\right) = \sum_{k=1}^{\infty} \dfrac{x^{k-1}}{k!} \implies \lim_{x \to 0} \dfrac{e^x - 1}{x} = 1$

19. $\dfrac{1 - \cos x}{x} = x^{-2}\left(1 - \displaystyle\sum_{k=0}^{\infty} (-1)^k \dfrac{x^{2k}}{(2k)!}\right) = \sum_{k=0}^{\infty} (-1)^k \dfrac{x^{2k}}{(2k+2)!} \implies \lim_{x \to 0} \dfrac{1 - \cos x}{x^2} = \dfrac{1}{2}$

21. $\dfrac{e^x - e^{-x}}{x} = 2\displaystyle\sum_{k=0}^{\infty} \dfrac{x^{2k}}{(2k+1)!} \implies \implies \lim_{x \to 0} \dfrac{e^x - e^{-x}}{x} = 2.$

23. $\dfrac{x - \arctan x}{x^3} = \displaystyle\sum_{k=0}^{\infty} (-1)^k \dfrac{x^{2k}}{2k+3} \implies \lim_{x \to 0} \dfrac{x - \arctan x}{x^3} = \dfrac{1}{3}$

25. $\dfrac{1 - \cos^2 x}{x} = \dfrac{\frac{1}{2} - \frac{1}{2}\cos(2x)}{x} = \displaystyle\sum_{k=1}^{\infty} (-1)^{k+1} \dfrac{(2x)^{2k-1}}{(2k)!} \implies \lim_{x \to 0} \dfrac{1 - \cos^2 x}{x} = 0$

27. $\sin\left(\sqrt{x}\right) = \displaystyle\sum_{k=0}^{\infty} (-1)^k \dfrac{x^{(2k+1)/2}}{(2k+1)!}$

29. $2^x = e^{x \ln 2} = \displaystyle\sum_{k=0}^{\infty} \dfrac{(x \ln 2)^k}{k!}$

31. $(x^2 - 1)\sin x = \left(x^2 - 1\right)\displaystyle\sum_{k=0}^{\infty} (-1)^k \dfrac{x^{2k+1}}{(2k+1)!} = -x + \displaystyle\sum_{k=1}^{\infty} (-1)^{k+1} \dfrac{\left((2k+1)(2k)+1\right)x^{2k+1}}{(2k+1)!} =$

$\displaystyle\sum_{k=0}^{\infty} (-1)^{k+1} \dfrac{(4k^2 + 2k + 1)x^{2k+1}}{(2k+1)!}$

33. $\cos^2 x = \frac{1}{2}\left(1 + \cos(2x)\right) = \dfrac{1}{2}\left(1 + \displaystyle\sum_{k=0}^{\infty} (-1)^k \dfrac{(2x)^k}{(2k)!}\right) = \dfrac{1}{2} + \displaystyle\sum_{k=0}^{\infty} (-1)^k \dfrac{2^{2k-1} x^{2k}}{(2k)!}$

35. $f(x) = \sin^3(x) = \frac{1}{4}\left(3 \sin x - \sin(3x)\right) = \dfrac{3}{4} \displaystyle\sum_{k=0}^{\infty} (-1)^k \dfrac{x^{2k+1}}{(2k+1)!} - \dfrac{1}{4}\displaystyle\sum_{k=0}^{\infty} (-1)^k \dfrac{(3x)^{2k+1}}{(2k+1)!} =$

$\displaystyle\sum_{k=1}^{\infty} (-1)^{k+1} \dfrac{3^{2k+1} - 3}{4 \cdot (2k+1)!} x^{2k+1}$

37. (a) Integrating term by term, $\dfrac{1}{1-x} = \displaystyle\sum_{k=0}^{\infty} x^k \implies -\ln|1-x| = \displaystyle\sum_{k=1}^{\infty} \dfrac{x^k}{k}.$

(b) The series converges on the interval $[-1, 1)$.

(c) When $x = 1/2$, part (a) implies that $-\ln(1/2) = \ln 2 = \displaystyle\sum_{k=1}^{\infty} \dfrac{1}{k\,2^k}$. Since the terms of this series are all

positive, the partial sums $S_N = \displaystyle\sum_{k=1}^{N} \dfrac{1}{k\,2^k}$ form an increasing sequence that is bounded above by the sum

of the series. Thus, $\ln 2 - S_N > 0$.

$$\ln 2 - S_N = \displaystyle\sum_{k=1}^{\infty} \dfrac{1}{k\,2^k} - \displaystyle\sum_{k=1}^{N} \dfrac{1}{k\,2^k} = \displaystyle\sum_{k=N+1}^{\infty} \dfrac{1}{k\,2^k} \le \dfrac{1}{N+1} \displaystyle\sum_{k=N+1}^{\infty} \dfrac{1}{2^k}$$

$$= \dfrac{1}{(N+1)2^{N+1}} \displaystyle\sum_{k=0}^{\infty} \dfrac{1}{2^k} = \dfrac{1}{(N+1)2^N}$$

39. The power series representations of each of the three functions is an alternating series if $x > 0$, so the following inequalities are valid if $0 < x < 1$: $x - x^2/2 < \ln(1+x) < x - x^2/2 + x^3/3$, $x - x^3/3! < \sin x < x$, and $x^2/2 - x^4/4! < 1 - \cos x < x^2/2$.

Since $(x - x^3/3!) - (x - x^2/2 + x^3/3) = x^2(1-x)/2 > 0$ if $0 < x < 1$, the lower bound on $\sin x$ is greater than the upper bound on $\ln(1+x)$ so $\ln(1+x) < \sin x$ if $0 < x < 1$. Also, since $(x - x^2/2) - x^2/2 = x(1-x) > 0$

if $0 < x < 1$, the lower bound on $\ln(1+x)$ is greater than the upper bound on $1 - \cos x$ so $1 - \cos x < \ln(1+x)$ if $0 < x < 1$.

41. (a) No. Since $\lim\limits_{n\to\infty} e^{-1/n} = 1 \neq 0$, the series diverges by the n-th term test.

(b) No. Using the power series representation of e^x and the alternating series theorem,

$$1 - e^{-1/n} > \frac{1}{n} - \frac{1}{2n^2}\frac{2n-1}{2n^2} \geq \frac{2n-n}{2n^2} = \frac{1}{2n}$$

for all $n \geq 1$. Therefore, $\sum\limits_{n=1}^{\infty}\left(1 - e^{-1/n}\right) \geq \frac{1}{2}\sum\limits_{n=1}^{\infty}\frac{1}{n}$.

43. (a) Despite first appearances, I is *not* a doubly improper integral: $\lim\limits_{x\to 0^+}\dfrac{xe^{-x}}{1-e^{-x}} = 1$. Therefore, since

$$I = \int_0^{\infty} \frac{xe^{-x}}{1-e^{-x}}\,dx = \int_0^1 \frac{xe^{-x}}{1-e^{-x}}\,dx + \int_1^{\infty} \frac{xe^{-x}}{1-e^{-x}}\,dx,$$

I converges if and only if the improper integral on right above converges. To show this, note that $1 - e^{-x} > 1/2$ for all $x \geq 1$. From this it follows that

$$\int_1^{\infty} \frac{xe^{-x}}{1-e^{-x}}\,dx < 2\int_1^{\infty} xe^{-x}\,dx = 4e^{-1}.$$

Therefore, I converges.

(b) If $u = 1 - e^{-x}$, $du = e^{-x}\,dx$, and $x = -\ln(1-u)$. Furthermore, $1 - e^{-0} = 0$ and $\lim\limits_{x\to\infty} 1 - e^{-x} = 1$, so

$$I = -\int_0^1 \frac{\ln(1-u)}{u}\,du.$$

(c) When $|u| < 1$, $\ln(1-u) = -\sum\limits_{k=1}^{\infty}\dfrac{u^k}{k}$. Therefore,

$$I = -\int_0^1 \frac{\ln(1-u)}{u}\,du = \int_0^1\left(\sum_{k=1}^{\infty}\frac{u^{k-1}}{k}\right)du = \sum_{k=1}^{\infty}\left(\int_0^1 \frac{u^{k-1}}{k}\,du\right) = \sum_{k=1}^{\infty}\frac{1}{k^2} = \frac{\pi^2}{6}.$$

45. (a) $\displaystyle\int e^{-x^2}\,dx = \int\left(\sum_{k=0}^{\infty}\frac{(-x^2)k}{k!}\right)dx = \sum_{k=0}^{\infty}\frac{(-1)^k}{(2k+1)\cdot k!}x^{2k+1}$

(b) The approximation $\displaystyle\int_0^1 e^{-x^2}\,dx \approx \sum_{k=0}^{3}\frac{(-1)^k}{(2k+1)\cdot k!} = \frac{26}{35}$ has the desired accuracy because $\dfrac{1}{9\cdot 4!} < 0.005$.

47. $\displaystyle\int_0^1 \sqrt{x}\sin x\,dx = \int_0^1\left(\sum_{k=0}^{\infty}(-1)^k\frac{x^{(4k+3)/2}}{(2k+1)!}\right)dx = \sum_{k=0}^{\infty}\frac{(-1)^k 2}{(4k+5)(2k+1)!} \approx$

$\displaystyle\sum_{k=0}^{2}\frac{(-1)^k 2}{(4k+5)(2k+1)!} = \frac{2557}{7020} \approx 0.36425$

49. $\dfrac{e^x}{1-x} = \left(\sum\limits_{k=0}^{\infty}\dfrac{x^k}{k!}\right)\left(\sum\limits_{k=0}^{\infty}x^k\right) = 1 + 2x + \dfrac{5}{2}x^2 + \dfrac{8}{3}x^3 + \cdots$

51. $e^{2x}\ln\left(1+x^3\right) = \left(\sum\limits_{k=0}^{\infty}\dfrac{(2x)^k}{k!}\right)\left(\sum\limits_{k=0}^{\infty}(-1)^{k+1}\dfrac{x^{3k}}{k}\right) = x^3 + 2x^4 + 2x^5 + \dfrac{5}{6}x^6 + \cdots$

53. $e^{\sin x} = \sum\limits_{k=0}^{\infty} \dfrac{(\sin x)^k}{k!} = 1 + \sin x + \dfrac{\sin^2 x}{2} + \dfrac{\sin^3 x}{3!} + \dfrac{\sin^4 x}{4!} + \cdots$

$\qquad = 1 + \left(x - x^3/3! + \cdots\right) + \dfrac{1}{2}\left(x - x^3/3! + \cdots\right)^2 + \dfrac{1}{3!}(x - \cdots)^3 + \dfrac{1}{4!}(x - \cdots)^4$

$\qquad = 1 + x + x^2/2 - x^4/8 + \cdots$

55. $-\cos x = \cos(x - \pi) = \sum\limits_{k=0}^{\infty}(-1)^k\dfrac{(x-\pi)^{2k}}{(2k)!} \implies \cos x = \sum\limits_{k=0}^{\infty}(-1)^{k+1}\dfrac{(x-\pi)^{2k}}{(2k)!}$

57. $\sum\limits_{k=1}^{\infty} kx^{k-1} = \left(\dfrac{1}{1-x}\right)' = \dfrac{1}{(1-x)^2}$

59. $\sum\limits_{k=1}^{\infty}(-1)^{k+1}x^k = 1 + \sum\limits_{k=0}^{\infty}(-1)^{k+1}x^k = 1 - \sum\limits_{k=0}^{\infty}(-1)^k x^k = 1 - \dfrac{1}{1+x} = \dfrac{x}{1+x}$

61.

63. $y = 2e^x = \sum\limits_{k=0}^{\infty}\dfrac{2x^k}{k!} \implies y' = \sum\limits_{k=0}^{\infty}\dfrac{2kx^{k-1}}{k!} = \sum\limits_{k=1}^{\infty}\dfrac{2x^{k-1}}{(k-1)!} = \sum\limits_{k=0}^{\infty}\dfrac{2x^k}{k!} = y$

65. $y = \sin x = \sum\limits_{k=0}^{\infty}(-1)^k\dfrac{x^{2k+1}}{(2k+1)!}$ and $y' = \sum\limits_{k=0}^{\infty}(-1)^k\dfrac{(2k+1)x^{2k}}{(2k+1)!} = \sum\limits_{k=0}^{\infty}(-1)^k\dfrac{x^{2k}}{(2k)!}$, so

$\qquad y'' = \sum\limits_{k=0}^{\infty}(-1)^k\dfrac{(2k)x^{2k-1}}{(2k)!} = \sum\limits_{k=1}^{\infty}(-1)^k\dfrac{x^{2k-1}}{(2k-1)!} = \sum\limits_{k=0}^{\infty}(-1)^{k+1}\dfrac{x^{2k+1}}{(2k+1)!} = -y.$

67. (a) $1 + \big(f(x)\big)^2 = 1 + \tan^2 x = \sec^2 x = f'(x)$

$\qquad$ (b) Since $f(0) = 0$, we assume that $f(x) = \sum\limits_{k=1}^{\infty} a_k x^k$. Inserting this power series into the identity from part (a): $a_1 + 2a_2 x + 3a_3 x^2 + \cdots = 1 + a_1^2 x^2 + 2a_1 a_2 x^3 + (2a_1 a_3 + a_2)x^4 + (2a_1 a_4 + 2a_2 a_3)x^5 + (2a_1 a_5 + 2a_2 a_4 + a_3^2)x^6 + \cdots$. Equating powers of x on both sides and solving, we find that $a_1 = 1$, $a_2 = 0$, $a_3 = 1/3$, $a_4 = 0$, $a_5 = 2/15$, $a_6 = 0$, and $a_7 = 17/315$. Thus, $\tan x = 1 + \dfrac{x^3}{3} + \dfrac{2x^5}{15} + \dfrac{17x^7}{315} + \cdots$.

69. Let $r = 1/2$.

71. $g(x) = \sqrt[3]{1 - x^2} \approx 1 - x^2/3 - x^4/9 - 5x^6/81$

73. $g(x) = \arcsin x = \displaystyle\int \dfrac{dx}{\sqrt{1-x^2}} = \int \left(1 + \dfrac{x^2}{2} + \dfrac{3x^4}{8} + \dfrac{5x^6}{16} + \cdots\right) dx = x + \dfrac{x^3}{6} + \dfrac{3x^5}{40} + \dfrac{5x^7}{112} + \cdots.$

11.7   Maclaurin and Taylor Series

1. (a) The Maclaurin series representation of f is the polynomial expression used to define f: $1 + 2x + 44x^2 - 12x^3 + x^4$.

   (b) $f(x) = 160 + 50(x-3) - 10(x-3)^2 + (x-3)^4$.

3. (a) $f'(x) = \sqrt{x}e^{-x}$, $f''(x) = e^{-x}\left(\frac{1}{2\sqrt{x}} - \sqrt{x}\right)$, and $f'''(x) = e^{-x}\left(\sqrt{x} - \frac{1}{\sqrt{x}} - \frac{1}{4x^{3/2}}\right)$. Therefore, $f(3) = 0$, $f'(3) = \sqrt{3}e^{-3}$, $f''(3) = -\frac{5\sqrt{3}}{6}$, and $f'''(3) = \frac{23\sqrt{3}}{36}$. It follows from Taylor's Theorem that

$$f(x) \approx \sqrt{3}e^{-3}(x-3) - \frac{5}{12}\sqrt{3}e^{-3}(x-3)^2 + \frac{23}{216}\sqrt{3}e^{-3}(x-3)^3.$$

   (b) If $3 \le x \le 3.5$, $\left|f^{(4)}(x)\right| \le 0.035$, so the approximation error is bounded by
   $$\frac{0.035 \cdot (0.5)^4}{4!} \approx 9.1146 \times 10^{-5}.$$

5. Since $f'(0) > 0$, the coefficient of x in the Maclaurin series representation of f must be positive; the coefficient of x in the series given is negative.

7. Let $f(x) = x^4 - 4x^3 + 5x$. Then $f(1) = 2$, $f'(1) = -3$, $f''(1) = -12$, $f'''(1) = 0$, $f^{(4)}(1) = 24$, and $f^{(k)}(1) = 0$ for all $k \ge 5$. Therefore, Taylor's Theorem implies that $f(x) = 2 - 3(x-1) - 6(x-1)^2 + (x-1)^4$. It follows that $a_0 = 2$, $a_1 = -3$, $a_2 = -6$, $a_3 = 0$, and $a_4 = 1$.

9. (a) No. The Maclaurin series representation of f is $f(x) = f(0) + f'(0)x + \frac{f''(0)}{2}x^2 + \cdots$. Since f is concave down at $x = 0$, the coefficient of x^2 in the Maclaurin series representation of f is negative.

   (b) Yes. $g''(0) = \frac{3}{4}\left(f'(0)\right)^2\left(g(0)\right)^5 - \frac{1}{2}f''(0)\left(g(0)\right)^3 > 0$.

11. (a) Using the fact that the Maclaurin series for $(1-u)^{-1}$ is $\displaystyle\sum_{k=0}^{\infty} u^k$, the MacLaurin series for f is $\displaystyle\sum_{k=0}^{\infty} x^{3k+1}$.

   (b) The interval of convergence of the power series for f is $(-1, 1)$.

   (c) Differentiating the power series for f term by term, $f''(x) = \displaystyle\sum_{k=1}^{\infty}(3k+1)(3k)x^{3k-1}$.

   (d) Integrating the power series for f term by term, $\displaystyle\int_0^x f(t)\,dt = \sum_{k=0}^{\infty}\frac{x^{3k+2}}{3k+2}$.

13. The Maclaurin series representation of f is $f(x) = \displaystyle\sum_{k=0}^{\infty}\frac{f^{(k)}(0)}{k!}x^k = \sum_{k=0}^{\infty}(-1)^k\frac{1}{(2k+2)!}x^{2k}$.

   Thus, $f^{(100)}(0)/100! = 1/102! \implies f^{(100)}(0) = 100!/102! = 1/102 \cdot 101 = 1/10302$.

15. (a) $f(x) = x^{-1}\sin x = x^{-1}\displaystyle\sum_{k=0}^{\infty}\frac{(-1)^k x^{2k+1}}{(2k+1)!} = \sum_{k=0}^{\infty}\frac{(-1)^k x^{2k}}{(2k+1)!}$

   (b) The power series in part (a) converges for values of x in the interval $(-\infty, \infty)$.

   (c) $f'''(x) = \displaystyle\sum_{k=2}^{\infty}\frac{(-1)^k(2k)(2k-1)(2k-2)x^{2k-3}}{(2k+1)!}$ for any $x \in (-\infty, \infty)$. Since $f'''(1)$ is represented by an alternating series, $\left|f'''(1) - \displaystyle\sum_{k=2}^{3}\frac{(-1)^k(2k)(2k-1)(2k-2)}{(2k+1)!}\right| < \frac{8\cdot7\cdot6}{9!} = \frac{1}{1080} < 0.005$;

   $f'''(1) \approx \dfrac{37}{210} \approx 0.17619$.

17. $K_{n+1} = e^x$ so $\left| e^x - P_n(x) \right| \le \dfrac{e^x \cdot |x|^{n+1}}{(n+1)!} \to 0$ as $n \to \infty$.

19. (a) If $\left| f^{(n)}(x) \right| \le n$ for all $n \ge 1$, then $K_{n+1} \le n+1$ and $|f(x) - P_n(x)| \le \dfrac{(n+1)|x|^{n+1}}{(n+1)!} \to 0$ as $n \to \infty$.

    (b) Yes, because $\displaystyle \lim_{n \to \infty} \frac{x^n}{n!} = 0$ for all x.

21. Start by writing $\cos x = \cos(x - \pi/3 + \pi/3) = \dfrac{1}{2}\cos(x - \pi/3) - \dfrac{\sqrt{3}}{2}\sin(x - \pi/3)$. Now, using the Maclaurin series for $\cos u$ and $\sin u$ (with $u = x - \pi/3$), the result is

$$\cos x = \frac{1}{2}\sum_{k=0}^{\infty} \frac{(-1)^k}{(2k)!}(x - \pi/3)^{2k} - \frac{\sqrt{3}}{2}\sum_{k=0}^{\infty} \frac{(-1)^k}{(2k+1)!}(x - \pi/3)^{2k+1}.$$

12.1 Differential equations: the basics

1. (a) If $y = Ce^x$ and $x = 0$, then $y = Ce^0 = C$. This means that the graph of $y = Ce^x$ has y-intercept C. The fact that $y'(0) = C$ means that the each graph has slope C at $x = 0$.

   (b) Each of the graphs has the appropriate y-intercept and slope at $x = 0$.

3. $y' = 3x^2$ so $xy' = x \cdot 3x^2 = 3x^3 = 3y$

5. (a) It sounds more or less reasonable. As the goal approaches, people's ardor to contribute seems likely to cool somewhat. On the other hand, as the goal becomes really near, some people might give money to get it over with.

   (b) As a DE, Neuman's Law of Cooling says $y' = k(y - 65)$.

   (c) "Neuman's Law of Cooling" suggests the analogy to Newton's Law of Cooling.

7. $y' = -1/x^2$ and $y'' = 2/x^3$ so $x^3 y'' + x^2 y' - xy = x$.

9. $y' = -C_1 \sin x + C_2 \cos x$ and $y'' = -C_1 \cos x - C_2 \sin x$. Thus, $y'' + y = (-C_1 \cos x - C_2 \sin x) + 1 + C_1 \cos x + C_2 \sin x = 1$.

11. If $y = x$ and $y = x^2$ are both solutions of the differential equation, the following conditions must hold: $g(x) + x = 0$ and $2f(x) + g(x) \cdot x + x^2 = 0$. Both conditions are satisfied if $f(x) = x^2/2$ and $g(x) = -x$.

12.2   Slope fields: solving DE's graphically

1. (a) The solution curves are "parallel" to each other in the sense that they differ from each other only in their *horizontal* position. Thus, e.g., all the curves have the same slope where $y = 2$.

    (b) It *does* appear that each of the five "upper" curves has the same slope when $y = 3$. Carefully draw a tangent line to any one of the curves at the appropriate point; measure its slope. The result should be 3 (or very close to 3).

    The answer *could* have been predicted in advance. The fact that each curve is a solution to the DE $y' = y$ means precisely that when $y = 3$, $y' = 3$, too.

    (c) At the level $y = -4$, each curve has slope -4. Again, this is exactly what the DE predicts.

    (d) All curves appear to be very nearly *horizontal* near $y = 0$. The only solution curve that actually touches the line $y = 0$ is the solution curve $y = 0$ itself. Appropriately, this curve has slope 0 everywhere.

3. (a) The straight line is $y = -t - 1$; it corresponds to $C = 0$.

    (b) The curve $y = 5e^t - t - 1$ passes through $(0, 4)$. One way to tell this is to solve the equation $5 = Ce^0 - 0 - 1$ for C.

    (c) At $(0, 4)$ the curve mentioned above has slope 4. This can be found (approximately) graphically by looking at slope, or symbolically by reading the DE.

    (d) The line $y + t = 0$ (aka $y = -t$) crosses each of the "upper" four solution curves at a *stationary point*, i.e., a point where the slope is zero. This happens because, as the DE demands, $y' = 0$ whenever $y + t = 0$.

    (e) The line $y + t = -3$ (aka $y = -t - 3$) crosses the "lower" four solution curves at four points. At each of these points, the solution curve in question has the same slope: -3. This is as it should be. As the DE requires at such points, $y' = y + t = -3$.

    (f) Any line with slope -1 crosses the solution curves at points of *equal slope*. As in the previous two parts, this is because the DE requires it.

    (g) Moving clockwise from upper left, the curves correspond to the C-values 500, 50, 5, 0.2, 0, -0.2, -5, -50, -500.

5. (a) In the slope field for $y' = 2t - 5$, all ticks at the same *horizontal position* are parallel. (That's because all such ticks have the same t-value.)

    (b) In the slope field for $y' = y$, all ticks at the same *vertical position* are parallel. (That's because all such ticks have the same y-value.)

7. (b) $y(x) = \sqrt{x^2 + 1}$

    (c) $y(x) = \pm\sqrt{x^2 - 1}$

9. (b) $y(x) = \dfrac{3}{2}e^{x^2} - \dfrac{1}{2}$

    (c) $y(x) = -\dfrac{1}{2e}e^{x^2} - \dfrac{1}{2} = -\dfrac{1}{2}\left(e^{x^2-1} + 1\right)$

12.3 Euler's method: solving DE's numerically

1. (a) A table helps keep track of results:

step	t	y'	y
0	0.00	0	0
1	0.25	0.2474	0
2	0.50	0.4794	0.06185
3	0.75	0.6816	0.1817
4	1.00	0.8415	0.3521

(b) The left rule with 4 subdivisions, applied to $I = \int_0^1 \sin(t)\, dt$, gives

$$L_4 = \frac{\sin(1/4)}{4} + \frac{\sin(1/2)}{4} + \frac{\sin(3/4)}{4} \approx 0.3521.$$

(c) The function $y(t) = 1 - \cos t$ solves the DE exactly. Thus, exactly, $y(1) = 1 - \cos 1 \approx 0.4597$.

(d) $\int_0^1 \sin(t)\, dt = -\cos t]_0^1 = -\cos 1 + \cos 0 = 1 - \cos 1 \approx 0.4597.$

(e) The error committed by L_4 is $|I - L_4| \approx 0.4597 - 0.3521 = 0.1076$.

(f) Here $y(t) = 1 - \cos t$, and $Y(t)$ is the function tabulated above. Thus we get:

t	0.00	0.25	0.50	0.75	1.00
$Y(t)$	0	0	0.0618	0.1817	0.3521
$y(t)$	0	0.0311	0.1224	0.2683	0.4597

(g) To plot $y(t)$ and $Y(t)$ on one pair of axes, we use the formula $1 - \cos t$ for y, and "connect the dots" for Y:

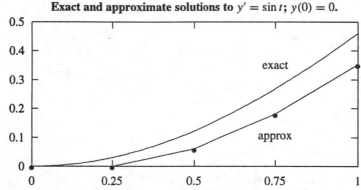

Exact and approximate solutions to $y' = \sin t$; $y(0) = 0$.

Notice that the two functions start out together, but spread apart as t increases.

3. (a) It's not hard to guess that the function $y(t) = 3t$ solves the IVP above. Thus $y(1) = 3$, $y(2) = 6$, $y(3) = 9$, $y(4) = 12$, $y(5) = 15$.

(b) In this case, Euler's method gives *exact* values; the Euler estimates commit *no* error.

(c) Euler's method pretends, in effect, that y' remains constant over small intervals. In this case, y' *is* constant, so Euler's method commits no error.

Euler's method will behave this way (i.e., commit *no* error) whenever y' is a constant function.

5. One does the "easy calculation" by checking explicitly that both the DE and the initial condition are satisfied. Since $y(t) = 70 + 120e^{-0.1t}$, it's easy to check that

$$y'(t) = -12e^{-0.1t} = -0.1 \cdot 120 \cdot e^{-0.1t}) = -0.1 \cdot (y - 70).$$

Thus, the DE does hold as advertised. Also, $y(0) = 70 + 120 = 190$, so the initial condition holds, too. Finally,

$$y(5) = 70 + 120e^{-0.1 \cdot 5} = 70 + 120e^{-0.5} \approx 142.78.$$

7. (a) Work quickly shows that given *initial* population 0, the population *remains* at 0. Nothing happens.

   (b) In biological terms, the situation is simply that without any initial breeding members a population can't grow. No parents; no children.

9. (a) $y(1) \approx -0.59374$

   (b) $y(1) \approx -0.65330$

   (c) $y'(t) = -e^t = (2 - e^t) - 2 = y(t) - 2$ and $y(0) = 1$

   (d) $y(1) = 2 - e \approx -0.71828$

11. (a) $y(0.8) \approx 2.6764$

   (b) $y'(t) = (1 - t)^{-2} = \left(y(t)\right)^2$ and $y(0) = 1$

   (c) The derivative is changing rapidly and is becoming large. Thus, very small steps are required to achieve an accurate result.

12.4   Separating variables: solving DE's symbolically

1. (a) $f(y) = 1/y$, $g(x) = x$

   (b) $F(y) = \ln y$, $G(x) = x^2/2$

   (c) Differentiating both sides of the equation $G(x) = F(y) + C$ with respect to x leads to the equation $G'(x) = F'(y)y'$ or $g(x) = f(y)y'$. Thus, if y is defined by the equation $G(x) = F(y) + C$, y also is a solution of the DE.

   (d) $G(x) = F(y) + K \implies x^2/2 = \ln y + K \implies e^{x^2/2} = e^K y \implies y = Ce^{x^2/2}$

Find a solution of each of the following separable differential equations.

3. $\arctan y = x + C$ or $y = \tan(x + C)$

5. $\arcsin y = x + C$ or $y = \sin(x + C)$

7. $-\ln|1 - y| = x + C$ or $y = 1 - Ae^{-x}$

9. $\ln(\ln y) = x + C$ or $y = e^{Ae^x}$

11. $y + y^3/3 = x^2/2 + C$

13. $-1/y = x^2/2 + C$ or $y = 2/(A - x^2)$

15. Separating variables in the DE gives

$$\frac{dP}{dt} = -0.00000556P(P - 10000) \implies \int \frac{dP}{P(P - 10000)} = -\int 0.00000556\, dt.$$

Integrating both sides gives

$$-\frac{\ln|P|}{10000} + \frac{\ln|P - 10000|}{10000} = -0.00000556t + C,$$

for some constant C. Setting $P = 1000$ and $t = 0$ gives

$$C = -\frac{\ln 1000}{10000} + \frac{\ln 9000}{10000} = \frac{\ln 9}{10000} \approx 0.0002197.$$

Thus, $P(t) = \dfrac{10000}{9e^{-0.0556t} + 1}$. Setting $t = 10$ and solving for P gives $P(10) \approx 1623$—not much different from what we've seen in other sections.

17. (a) We need to check (1) that $y(t) = (T_0 - T_r)e^{-0.1t} + T_r$ solves the DE $y' = -0.1(y - T_r)$; and (2) that $y(0) = T_0$. The latter is easy: $y(0) = (T_0 - T_r)e^0 + T_r = T_0$, as claimed.

   To check (1), we calculate both sides of the DE and see that they agree:

$$y(t)' = (T_0 - T_r) \cdot e^{-0.1t} \cdot (-0.1);$$
$$-0.1(y - T_r) = (-0.1) \cdot (T_0 - T_r)e^{-0.1t}.$$

   (b) If $T_0 = 200$ and $y(10) = 100$, then

$$y(10) = (200 - T_r)e^{-1} + T_r = 100 \implies T_r = -\frac{200\,e^{-1} - 100}{-e^{-1} + 1} \approx 41.802.$$

   Given this value of T_r,

$$y(20) \approx (200 - 41.802)e^{-2} + 41.802 \approx 63.212.$$

   In coffee language, this means that the coffee temperatures were 200 degrees, 100 degrees, and about 63 degrees at times 0, 10, and 20, respectively.

(c) If $T_r = 80$ and $y(10) = 120$, then

$$y(10) = (T_0 - 80)e^{-1} + 80 = 120 \implies T_0 = -\left(-\frac{80}{e} - 40\right)e \approx 188.731.$$

With this value for T_0, we have

$$y(20) \approx (188.731 - 80)e^{-2} + 80 \approx 94.715.$$

In coffee language, this means that the coffee temperatures were around 189 degrees, 120 degrees, and 95 degrees at times 0, 10, and 20, respectively.

(d) We can use the information $y(10) = 100$ and $y(20) = 80$ to solve for *both* constants T_0 and T_r, as follows:

$$y(10) = 100 \implies 100 = (T_0 - T_r) \cdot e^{-1} + T_r;$$
$$y(20) = 80 \implies 80 = (T_0 - T_r) \cdot e^{-2} + T_r.$$

Solving these two equations gives $Tr \approx 68.36$, $T_0 \approx 154.36$. In other words, the coffee was about 154 degrees at time 0; room temperature is about 68 degrees. When $t = 40$, the temperature will be $y(40) = (154.36 - 68.36)e^{-4} + 68.36 \approx 69.94$ degrees.

(e) Since $\lim\limits_{t \to \infty} e^{-0.1t} = 0$,

$$\lim_{t \to \infty} y(t) = \lim_{t \to \infty} (T_0 - T_r)e^{-0.1t} + T_r = T_r.$$

In coffee language, this means (as experience suggests!) that in the long run coffee cools to room temperature.

19. The point is that $e^{-x} \to 0$ as $x \to \infty$. Now if k and C are positive, $Ckt \to \infty$ as $t \to \infty$. Therefore $e^{-Ckt} \to 0$ as $t \to \infty$. Thus

$$\lim_{t \to \infty} P(t) = \lim_{t \to \infty} \frac{C}{Ke^{-Ckt} + 1} = \frac{C}{K \cdot 0 + 1} = C.$$

In biological terms, this means that in the long run, the population tends toward C, its carrying capacity.

21. (a) Differentiating the right side of $P' = kP(C - P)$ with respect to P gives

$$\frac{d}{dP}(kP(C - P)) = \frac{d}{dP}\left(kCP - kP^2\right) = kC - 2KP = 0 \iff C = \frac{P}{2}.$$

Thus P' has its maximum value where $P = C/2$, as claimed. (In rumor language: The rumor spreads fastest when *half* the people know it.)

(b) Let's solve $P(t) = 500$ for t (check the steps):

$$P(t) = \frac{1000}{49e^{-0.947t} + 1} = 500 \implies 2 = 49e^{-0.947t} + 1 \implies t = \frac{\ln 49}{0.947} \approx 4.11.$$

(In rumor language: After 4.11 days, half the people know the rumor.) (Notice that one could also have read this from the graph.)

23. For each part, we know two things:

$$P'(t) = \frac{K}{1 + t}P(1000 - P) \quad \text{and} \quad P(t) = \frac{1000}{d(1 + t)^{-1000K} + 1}.$$

The point is to use given information to find explicit values for K and d.

(a) If $P(0) = 100$ and $P'(0) = 50$, then

$$P(0) = \frac{1000}{d+1} = 100 \implies d = 9;$$

$$P'(0) = K \cdot 100 \cdot 900 = 50 \implies K = \frac{1}{1800}.$$

With these values for K and d, we have

$$P(t) = \frac{1000}{9(1+t)^{-1000/1800} + 1} = \frac{1000}{9(1+t)^{-5/9} + 1}.$$

(b) If $P(0) = 50$ and $P'(0) = 100$, then

$$P(0) = \frac{1000}{d+1} = 50 \implies d = 19;$$

$$P'(0) = K \cdot 100 \cdot 900 = 100 \implies K = \frac{1}{900}.$$

Therefore

$$P(t) = \frac{1000}{19(1+t)^{-1000/900} + 1} = \frac{1000}{19(1+t)^{-10/9} + 1}.$$

(c) If $P(0) = 800$ and $P'(0) = 50$, then (as above) $d = 1/4, k = 1/3200$, so $P(t) = \dfrac{4000}{(1+t)^{-5/16} + 4}.$

(d) If $P(0) = 800$ and $P'(0) = 100$, then (as above) $d = 1/4, k = 1/1600$, so $P(t) = \dfrac{4000}{(1+t)^{-5/8} + 4}.$

(e) Here are graphs, machine drawn:

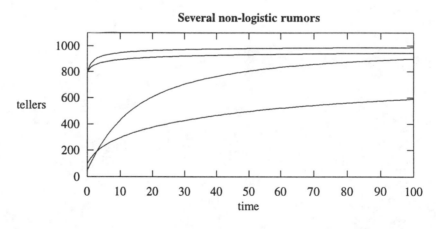

25. Amy drinks the hotter coffee. Let $A(t)$ be the temperature of Amy's coffee and $J(t)$ be the temperature of Joan's coffee. Also, let $J(0)$ be the initial temperature of both cups of coffee (before any cream is added) and $A(0)$ be the temperature of Amy's coffee after the cream is added. The effect of adding the cream to a cup of coffee is to reduce the temperature of the coffee by by $J(0) - A(0)$. Now, because of Newton's law of cooling, we know that $J(10) - A(10) < J(0) - A(0)$ so $J(10) - (J(0) - A(0)) < A(10)$.

NOTE: This argument is only qualitatively correct. A more careful analysis requires taking the relative amounts of the coffee and the cream into account.

27. Let $f(x) = 2x$ and $g(x) = x$. Then $F(x) = x^2$ and $\displaystyle\int_0^x e^{F(t)} g(t)\, dx = e^{x^2}/2 - 1/2$. Therefore, the solution of the IVP is $y(x) = 1/2 - e^{-x^2}/2$.

Use part (b) of Exercise 26 to find the solution of each of the IVPs in Exercises 28–31.

29. Here $f(x) = x$ and $g(x) = e^{2x}$. Therefore, the solution of the IVP is $y(x) = e^{2x}/3 + (2e - e^3/3)e^{-x}$.

31. Start by rewriting the DE in the form $y' + \dfrac{2}{x}y = \dfrac{\sin x}{x}$. Then $f(x) = 2/x$ and $g(x) = (\sin x)/x$, so $F(x) = 2\ln x$, $e^{F(x)} = x^2$, and $\displaystyle\int_{\pi/2}^{x} e^{F(t)}g(t)\,dt = \sin x - x\cos x - 1$. Therefore, the solution of the IVP is

$$y(x) = \frac{\pi^2/4 + \sin x - x\cos x - 1}{x^2}.$$

33. (a)

   (b) The DE is a Bernoulli equation with $n = -1$. The solution is $y(x) = x\sqrt{C + 2\ln x}$.

13.1   Polar coordinates and polar curves

1. $(\sqrt{2}, \pi/4)$; $(\sqrt{2}, -7\pi/4)$; $(-\sqrt{2}, -3\pi/4)$; $(-\sqrt{2}, 5\pi/4)$

3. $(2, \pi/3)$; $(2, -5\pi/3)$; $(-2, 4\pi/3)$; $(-2, -2\pi/3)$

5. $(\pi, 0)$; $(-\pi, \pi)$; $(\pi, 2\pi)$

7. $(\sqrt{5}, 1.107)$; $(\sqrt{5}, -5.176)$; $(-\sqrt{5}, 4.249)$

9. $(\sqrt{17}, 1.326)$; $(\sqrt{17}, -4.957)$; $(-\sqrt{17}, 4.467)$

11. $(0.596, 0.931)$; $(0.596, -5.352)$; $(-0.596, 4.072)$

13. $(\sqrt{2}, \sqrt{2})$

15. $(\sqrt{3}/2, 1/2)$

17. $(a, 0)$

19. $(0.540, 0.841)$

21. $(-0.832, 1.82)$

23. $(\sqrt{2}/2, \sqrt{2}/2)$

25. Points of the form $(a, 0)$ where $a \in \mathbb{R}$ have this property.

27. (a)

θ	0	$\frac{\pi}{6}$	$\frac{\pi}{3}$	$\frac{\pi}{2}$	$\frac{2\pi}{3}$	$\frac{5\pi}{6}$	π	$\frac{7\pi}{6}$	$\frac{4\pi}{3}$	$\frac{3\pi}{2}$	$\frac{5\pi}{3}$	$\frac{11\pi}{6}$	2π
r	2	1.866	1.5	1	0.5	0.134	0	0.134	0.5	1	1.5	1.866	2

The cardioid is symmetric with respect to the x-axis.

41. (a) The polar points $(1, 0)$, $(1, 2\pi)$, and $(-1, \pi)$ all represent the Cartesian point $(1, 0)$. (Use $x = r\cos\theta$ and $y = r\sin\theta$.)

(b) The polar points $(-1, \pi/4)$, $(-1, 9\pi/4)$, and $(1, 5\pi/4)$ all represent the Cartesian point $(-\sqrt{2}/2, -\sqrt{2}/2)$.

(c) $(\sqrt{2}, \pi/4 + 2k\pi)$ and $(-\sqrt{2}, \pi/4 + (2k - 1)\pi)$

43. $x^2 + y^2 = 16$

45. $r = 2\sin\theta \implies r^2 = 2r\sin\theta \implies x^2 + y^2 = 2y$

47. $r = 4\csc\theta$

49. $r = 2\cos\theta$

13.2   Calculus in polar coordinates

1. The results are straightforward applications of the product rule: $(f \cdot g)' = f' \cdot g + f \cdot g'$.

3. (b) $\dfrac{dy}{dx} = \dfrac{\sin\theta + \theta\cos\theta}{\cos\theta - \theta\sin\theta}$. Thus, the spiral has a horizontal tangent line wherever $\theta = -\tan\theta$ and a vertical tangent line wherever $\theta = \cot\theta$.

   (c) The polar point $(1, 1)$ is the point $(\cos 1, \sin 1)$ in Cartesian coordinates. The slope of the tangent line at the polar point $(1, 1)$ is $m = (\sin 1 + \cos 1)/(\cos 1 - \sin 1) \approx -4.588$. Thus, the equation of the desired tangent line is $y = m(x - \cos 1) + \sin 1 \approx -4.59(x - 0.54) + 0.84$.

5. (b) $\dfrac{dy}{dx} = \dfrac{-a\sin^2\theta + \cos\theta + a\cos^2\theta}{-2a\sin\theta\cos\theta - \sin\theta}$. The limaçon has a vertical tangent whenever the denominator in this expression is zero (i.e., when $\theta = 0$ or $\cos\theta = -1/2a$). Thus, there will be three vertical tangent lines if and only if $|a| \le 1/2$.

7. area $= 9\pi/2$

9. area $= \beta/2$

11. area $= \dfrac{1}{2}\displaystyle\int_{7\pi/6}^{11\pi/6} f(\theta)^2\, d\theta = \pi - 3\sqrt{3}/2$

13. area $= m/2$

15. (a) One leaf lies between $\theta = -\pi/2n$ and $\theta = \pi/2n$. The area of this leaf is $\dfrac{1}{2}\displaystyle\int_{-\pi/2n}^{\pi/2n} \cos^2(n\theta)\, d\theta = \pi/4n$.

    (b) The area of all n leaves is $\pi/4$ (i.e., one-fourth of the area of the circle $r = 1$).

17. area $= \left(e^{4\pi} - 1\right)/4$

19. area $= \dfrac{1}{2}\displaystyle\int_{-\pi/3}^{\pi/3} d\theta - \dfrac{1}{2}\displaystyle\int_{-\pi/3}^{\pi/3}\left(\tfrac{1}{2}\sec\theta\right)^2 d\theta = \dfrac{\pi}{3} - \dfrac{\sqrt{3}}{4}$

21. $x = 2\cos t$, $y = 2\sin t$, $0 \le t \le 2\pi$. The graph is a circle of radius 2 with center at the origin.

23. $x = 1$, $y = \tan t$, $-\pi/4 \le t \le \pi/4$. The graph is the vertical line segment from $(1, -1)$ to $(1, 1)$.

25. $x = t\cos t$, $y = t\sin t$, $0 \le t \le 2\pi$. The graph is one loop of a spiral.

27. $x = \cos(2t)\cos t$, $y = \cos(2t)\sin t$, $0 \le t \le 2\pi$. The graph is the 4-leaf rose shown at the beginning of the section.

29. $\left(\dfrac{dy}{d\theta}\right)^2 + \left(\dfrac{dx}{d\theta}\right)^2 = \left(f'(\theta)\sin\theta + f(\theta)\cos\theta\right)^2 + \left(f'(\theta)\cos\theta - f(\theta)\sin\theta\right)^2$ which implies that $dx/d\theta$

$= \left(f'(\theta)^2\sin^2\theta + 2f(\theta)f'(\theta)\sin\theta\cos\theta + f(\theta)^2\cos^2\theta\right) +$
$\left(f'(\theta)^2\cos^2\theta - 2f(\theta)f'(\theta)\sin\theta\cos\theta + f(\theta)^2\sin^2\theta\right)$

$= f'(\theta)^2 + f(\theta)^2 > 0$

and $dy/d\theta$ are not both simultaneously zero.

14.1   Three-dimensional space

1. We said in this section that the graph of $y = x^2$ is a cylinder in xyz-space, unrestricted in the z-direction. (See the picture on page 386.)

   (a) Plot the equation $z = y^2$ in xyz-space. What is the unrestricted direction?   [HINT: Start in yz-space.]

   (b) Plot the equation $z = x^2$ in xyz-space. What is the unrestricted direction?

2. Plot the equation $x^2 + y^2 = 1$, first in xy-space, then in xyz-space. (The second graph should resemble a vertical pipe—a cylinder in the everyday sense of the word.)

3. In each part, plot the given equation in xyz-space (a rough sketch is fine), then describe the graph in words.

   (a) $y^2 - z^2 = 0$

   (b) $y^2 + z^2 = 0$

   (c) $y^2 + z = -1$

   (d) $y^2 + z^2 = -1$ [HINT: Does the graph contain any points?]

4. Find an equation in x, y, and z for the graph in xyz-space of

   (a) a sphere of radius 2, centered at the origin.

   (b) a sphere of radius 1, centered at $(1, 1, 1)$.

   (c) a circular cylinder of radius 1, centered along the y-axis.

   (d) a circular cylinder of radius 2, centered along the z-axis.

   (e) a cylindrical surface that resembles an ocean, with waves rolling in the x-direction (see Example 4, page 386).

5. The equation $z = 3$ omits *two* variables. Therefore, its graph in xyz-space should be a cylinder in both the x-direction and the y-direction. Is it? What is the trace of the graph in each of the coordinate planes?

6. The graph of $x^2 + y^2 - 6y + z^2 - 4z = 0$ is a sphere. Find the center and radius; then draw the sphere.

7. Consider the unit sphere S, with equation $x^2 + y^2 + z^2 = 1$. If we set $z = 0$ in this equation, we get $x^2 + y^2 + z^2 = x^2 + y^2 = 1$. This means, geometrically, that S intersects the xy-plane in the unit circle $x^2 + y^2 = 1$. In mathematical language, the unit circle is the trace of S in the xy-plane. (To put it another way, the unit circle is the "equator" of the unit sphere.)

   (a) Set $x = 0$ in the original equation to find the equation of the intersection of S and the yz-plane.

   (b) What is the intersection of S and the xz-plane? Describe the answer geometrically.

   (c) Use the results in parts (a) and (b) to sketch the part of S that lies in the first octant.   [HINTS: First draw a set of coordinate axes. Then draw the traces of S in each of the three coordinate planes.]

   (d) Set $z = 1/2$ in the original equation to show that S intersects the plane $z = 1/2$ in the circle of radius $\sqrt{3}/2 \approx 0.87$, centered at $(0, 0)$.

    (e) What is the trace of S in the plane $z = 0.9$? In the plane $z = 1$? In the plane $z = 2$? Explain your answers.

8. The distance formula in xyz-space can be thought of as just another instance of the Pythagorean rule for right triangles. (The square of the hypotenuse is the sum of the squares of the sides.) This exercise illustrates why.

    (a) Plot and label the points $O(0, 0, 0)$, $P(1, 0, 0)$, $Q(1, 2, 0)$, and $R(1, 2, 3)$ in an xyz-coordinate system. Observe that the triangles $\triangle OPQ$ and $\triangle OQR$ are both right triangles. Mark the sides OP, PQ, and QR with their lengths. (The lengths should be obvious from the picture.)

    (b) Use the Pythagorean rule (not the distance formula) on the triangle $\triangle OPQ$ to find the length of OQ.

    (c) Use the Pythagorean rule (not the distance formula) on the triangle $\triangle OQR$ to find the length of OR.

    (d) For comparison, use the distance formula to compute the lengths of OQ and OR.

9. Any reasonable formula for distance should satisfy some commonsense requirements. For example, the distance $d(P, P)$ from any point P to *itself* should certainly be zero. So it is. If $P(x, y, z)$ is any point, then the distance formula says

$$d(P, P) = \sqrt{(x - x)^2 + (y - y)^2 + (z - z)^2} = 0.$$

In the same spirit, use the distance formula to verify the following commonsense properties. Throughout, use the points $P(x, y, z)$ and $Q(a, b, c)$.

    (a) If $P \neq Q$, then $d(P, Q) > 0$.

    (b) $d(P, Q) = d(Q, P)$.

    (c) If M is the midpoint of P and Q, then $d(P, M) = d(M, Q) = d(P, Q)/2$.

10. In this section we said that xyz-space contains 8 different octants. List 8 points, all with coordinates ± 1, one in each octant. Draw a picture showing all 8 points.

11. Any line in the xy-plane (even a vertical line) has an equation of the form $Ax + By = C$ for some constants A, B, and C. For each of the following lines, state an equation in this form.

    (a) The line $y = 3x + 5$

    (b) The line $y = mx + b$, where m and b are any constants

    (c) The horizontal line through $(2, 3)$

    (d) The vertical line through $(2, 3)$

12. Consider the linear equation $Ax + By = C$ and its graph (a line) in the xy-plane. Here A, B, and C are constants, and we assume that A and B aren't both zero.

    (a) We assumed that A and B aren't both zero. What goes wrong if $A = B = 0$?

    (b) Find the slope of the line $Ax + By = C$. Which lines have undefined slope?

(c) Find the y-intercept of the line $Ax + By = C$. Which lines have no y-intercept?

(d) Find the x-intercept of the line $Ax + By = C$. Which lines have no x-intercept?

13. Consider the linear equation $Ax + By + Cz = D$ and its graph (a plane) in xyz-space. Here A, B, C, and D are all constants, and we assume that A, B, and C aren't all zero.

(a) We assumed that A, B, and C aren't all zero. What goes wrong if $A = B = C = 0$?

(b) Find (if possible) an x-intercept of the plane $Ax + By + Cz = D$. (Set $y = 0$ and $z = 0$, and then solve for x.) Give an example of a plane with no x-intercept.

(c) Find (if possible) a z-intercept of the plane $Ax + By + Cz = D$. Give an example of a plane with no z-intercept.

14. The linear equation $x + 2y + 3z = 3$ defines a plane p in xyz-space. (See the picture on page 384.)

(a) Find the equation of the trace of p in the xz-plane. Where does the trace intercept the x- and z-axes?

(b) Find the equation of the trace of p in the yz-plane. Where does the trace intercept the y- and z-axes?

(c) Find the equation of the trace of p in the plane $x = 1$.

15. Consider the plane p with equation $4x + 2y + z = 4$.

(a) Find the x-, y-, and z-intercepts of p. Use them to draw p in the first octant.

(b) Find the traces of p in each of the three coordinate planes. How do your answers appear in the picture in part (a).

16. We said in this section that, as a rule, a line in the xy-plane intercepts both coordinate axes, and a plane in xyz-space intercepts all three coordinate axes. But exceptions are possible, as this exercise explores.

(a) Give an example of a line in the xy-plane that intercepts the x-axis but not the y-axis. Write an equation for your line in the form $ax + by = c$.

(b) Consider the plane $x = 1$ in xyz-space. Find all possible intercepts with the three coordinate axes.

(c) Consider the plane $x + 2y = 1$ in xyz-space. Find all possible intercepts with the three coordinate axes.

(d) Give the equation of a plane in xyz-space that intersects the the y-axis and the z-axis but not the x-axis.

17. Suppose that $P_1(x_1, y_1, z_1)$ and $P_2(x_2, y_2, z_2)$ both lie on the plane with equation $Ax + By + Cz = D$. Show that the midpoint of P_1 and P_2 also lies on this plane.

18. Here's another way of defining orientation of a system of axes. The system is right-handed if it's possible to point (all at once!) the index finger, the middle finger, and the thumb of the right hand along the positive x-, y-, and z-axes. Otherwise the system is left-handed. Use this definition to draw two more systems (different from those on page 387), one with each orientation.

19. Two systems of positive x-, y-, and z-axes have the same orientation if one can be "rotated into the other," i.e., if one system of axes can be picked up and turned around to "fit onto" the other. Explain why this can be done with the left and middle systems on page 387, but not with the left and right systems.

14.2 Functions of several variables

1. Find the range and domain of each function.

    (a) $g(x, y) = x^2 + y^2$

    (b) $h(x, y) = x^2 + y^2 + 3$

    (c) $j(x, y) = 1/(x^2 + y^2)$

    (d) $k(x, y) = x^2 - y^2$

    (e) $m(x, y) = \sqrt{1 - x^2 - y^2}$

2. Let $f(x, y) = y - x^2$ and let $g(x, y) = x - y^2$.

    (a) In the rectangle $[-3, 3] \times [-3, 3]$, draw and label the level curves of f that correspond to $z = -3, z = -2, \dots, z = 2$, and $z = 3$. What is the shape of each level curve?

    (b) In the rectangle $[-3, 3] \times [-3, 3]$, draw and label the level curves of g that correspond to $z = -3, z = -2, \dots, z = 2$, and $z = 3$. What is the shape of each level curve?

    (c) How are the results of parts (a) and (b) similar? How are they different?

    (d) Use technology to plot the graphs $z = f(x, y)$ and $z = g(x, y)$, for (x, y) in $[-3, 3] \times [-3, 3]$. Describe briefly, in words, how the two graphs are related to each other.

3. Let $f(x, y) = x^2 + y^2$ and let $g(x, y) = x^2 + y^2 + 1$.

    (a) In the rectangle $[-3, 3] \times [-3, 3]$, draw and label the level curves of f that correspond to $z = 0, z = 2, z = 4, z = 6$, and $z = 8$. What is the shape of each level curve?

    (b) In the rectangle $[-3, 3] \times [-3, 3]$, draw and label the level curves of g that correspond to $z = 1, z = 3, z = 5, z = 7$, and $z = 9$. What is the shape of each level curve?

    (c) How are the results of parts (a) and (b) similar? How are they different?

    (d) Use technology to plot the graphs $z = f(x, y)$ and $z = g(x, y)$, for (x, y) in $[-3, 3] \times [-3, 3]$. Describe briefly, in words, how the two graphs are related to each other.

4. Let f and g be the linear functions $f(x, y) = 2x - 3y$ and $g(x, y) = -2x + 3y$.

    (a) In the rectangle $[-3, 3] \times [-3, 3]$, draw and label the level curves of f that correspond to $z = -5, z = -3, z = -1, z = 1, z = 3$, and $z = 5$. What is the shape of each level curve?

    (b) In the rectangle $[-3, 3] \times [-3, 3]$, draw and label the level curves of g that correspond to $z = -5, z = -3, z = -1, z = 1, z = 3$, and $z = 5$. What is the shape of each level curve?

    (c) How are the results of parts (a) and (b) similar? How are they different?

    (d) What special properties do the level curves of a *linear* function have?

    (e) Use technology to plot the graphs $z = f(x, y)$ and $z = g(x, y)$, for (x, y) in $[-3, 3] \times [-3, 3]$. Describe briefly, in words, how the two graphs are related to each other.

5. Let f and g be the functions $f(x, y) = 2 + x^2$ and $g(x, y) = 2 + y^2$.

   (a) In the rectangle $[-3, 3] \times [-3, 3]$, draw and label the level curves of f that correspond to $z = 0$, $z = 2$, $z = 4$, $z = 6$, and $z = 8$. What is the shape of each level curve?

   (b) In the rectangle $[-3, 3] \times [-3, 3]$, draw and label the level curves of g that correspond to $z = 0$, $z = 2$, $z = 4$, $z = 6$, and $z = 8$. What is the shape of each level curve?

   (c) How are the results of parts (a) and (b) similar? How are they different?

   (d) Use technology to plot the graphs $z = f(x, y)$ and $z = g(x, y)$, for (x, y) in $[-3, 3] \times [-3, 3]$. Describe briefly, in words, how the two graphs are related to each other.

   (e) The graphs of f and g are both "cylinders" (in the sense we defined in Section 14.1). How do the contour maps of f and g reflect this fact?

6. Imagine a map of the United States in the usual position. The positive x-direction is east, and the positive y-direction is north. Suppose that the units of x and y are miles and that Los Angeles, California, has coordinates $(0, 0)$. (Several approximations are involved here. The earth's surface is not flat, and Los Angeles occupies more than a single point.) Let $T(x, y)$ be the temperature, in degrees Celsius, at the location (x, y) at noon, Central Standard Time, on January 1, 1996.

   (a) What does it mean in weather language to say that $T(0, 0) = 15$.

   (b) What do the level curves of T mean in weather language? As a rule, would you expect, level curves of T to run north and south or east and west? Why?

   (c) International Falls, Minnesota, is about 1400 miles east and 1100 miles north of Los Angeles. The noon temperature in International Falls on January 1, 1996, was -15 degrees Celsius. What does this mean about $T(x, y)$?

   (d) Suppose that International Falls was the coldest spot in the country at the time in question. How would you expect the level curves to look near International Falls?

7. For any point (x, y) in the xy-plane, let $f(x, y)$ be the distance from (x, y) to the origin. Then f has the formula $f(x, y) = \sqrt{x^2 + y^2}$.

   (a) Find the range and domain of f.

   (b) Plot f. Describe the graph in words.

   (c) Draw the level curve of f that passes through $(3, 4)$.

   (d) All level curves of f have the same shape. What is it?

8. For any point (x, y) in the xy-plane, let $f(x, y)$ be the distance from (x, y) to the line $x = 1$.

   (a) Find a formula for $f(x, y)$.

   (b) Plot f. Describe the graph in words.

   (c) Draw the level curve that passes through $(3, 4)$.

   (d) All level curves of f have the same shape. What is it?

9. Here are some values of a linear function $L(x, y)$. (No explicit symbolic formula for L is given.) Use the table to answer the following questions.

Values of $L(x, y)$

$y \backslash x$	-3	-2	-1	0	1	2	3
3	-15	-12	-9	-6	-3	0	3
2	-13	-10	-7	-4	-1	2	5
1	-11	-8	-5	-2	1	4	7
0	-9	-6	-3	0	3	6	9
-1	-7	-4	-1	2	5	8	11
-2	-5	-2	1	4	7	10	13
-3	-3	0	3	6	9	12	15

(a) All the level curves of L are straight lines. Using this fact, draw (all in the rectangle $[-3, 3] \times [-3, 3]$) and label the level curves $z = -12$, $z = -8$, $z = -4$, $z = 0$, $z = 4$, $z = 8$, $z = 12$.

(b) Find an equation in x and y for the level line $z = 0$.

(c) Because L is a linear function, its formula has the form $L(x, y) = a + bx + cy$, for some constants a, b, and c. Find numerical values for a, b, and c. [HINT: The table says that $L(0, 0) = 0$. Therefore, $L(0, 0) = a + b \cdot 0 + c \cdot 0 = 0$, so $a = 0$. Use similar reasoning to find values for b and c.]

(d) Use technology to plot $L(x, y)$ over the rectangle $[-3, 3] \times [-3, 3]$. Is the shape of the graph consistent with the level curves you plotted in part (a)?

14.3   Partial derivatives

1. For each of the following functions, find the partial derivative with respect to each variable.

   (a) $f(x, y) = x^2 - y^2$

   (b) $f(x, y) = x^2 y^2$

   (c) $f(x, y) = \dfrac{x^2}{y^2}$

   (d) $f(x, y) = \cos(xy)$

   (e) $f(x, y) = \cos(x)\cos(y)$

   (f) $f(x, y) = \dfrac{\cos(x)}{\cos(y)}$

   (g) $f(x, y, z) = xy^2 z^3$

   (h) $f(x, y, z) = \cos(xyz)$

2. Let $f(x) = x^2$ and let $x_0 = 3$.

   (a) Let L be the linear approximation to f at x_0. Show that $L(x) = 9 + 6(x - 3)$.

   (b) Plot L and f on the same axes. (Choose your own plotting window.) Supposedly, "L linearly approximates f near x_0." How do the graphs illustrate this?

   (c) Find an interval $a \le x \le b$ on which $|f(x) - L(x)| < 0.01$. (On this interval, $L(x)$ approximates $f(x)$ to within 0.01.)   [HINT: This can be done either graphically, by zooming, or symbolically, by solving inequalities.]

3. Redo Exercise 2, but work with $f(x) = \sqrt{x}$ and $x_0 = 9$.

4. Let $f(x) = x^2$.

   (a) Since $f'(x) = 2x$, $f'(3) = 6$. What does this mean about the graph of f? (Use a graph of f for $2.5 \le x \le 3.5$ to illustrate your answer.)

   (b) Plot f over the interval $2.5 \le x \le 3.5$. (It's OK to use a calculator or computer, but copy the graph onto paper.) On your graph, draw the secant lines from $x = 3$ to $x = 3.5$ and from $x = 3$ to $x = 3.1$. Find their slopes.

   (c) Find the average rate of change $\Delta y / \Delta x$ of f over the intervals $[3, 3.5]$ and $[3, 3.1]$.   [HINT: The answers should be familiar from part (b).]

   (d) Find the limit $\lim\limits_{h \to 0} \dfrac{f(3 + h) - f(3)}{h}$. Interpret the answer as a derivative. Does the answer agree with other information already given?

5. Redo Exercise 4, but use $f(x) = x^2 - x$.

6. Here are some values of a function $g(x, y)$. (No explicit symbolic formula for

g is given.) Use the table to answer the following questions.

Values of $g(x, y)$									
$y \backslash x$	−0.0100	0.0000	0.0100	0.0200	...	0.9900	1.0000	1.0100	1.0200
1.02	2.0603	2.0604	2.0603	2.0600	...	1.0803	1.0604	1.0403	1.0200
1.01	2.0300	2.0301	2.0300	2.0297	...	1.0500	1.0301	1.0100	0.9897
1.00	1.9999	2.0000	1.9999	1.9996	...	1.0199	1.0000	0.9799	0.9596
0.99	1.9700	1.9701	1.9700	1.9697	...	0.9900	0.9701	0.9500	0.9297
⋮	⋮	⋮	⋮	⋮	⋮	⋮	⋮	⋮	⋮
0.02	0.0203	0.0204	0.0203	0.0200	...	−0.9597	−0.9796	−0.9997	−1.0200
0.01	0.0100	0.0101	0.0100	0.0097	...	−0.9700	−0.9899	−1.0100	−1.0303
0.00	−0.0001	0.0000	−0.0001	−0.0004	...	−0.9801	−1.0000	−1.0201	−1.0404
−0.01	−0.0100	−0.0099	−0.0100	−0.0103	...	−0.9900	−1.0099	−1.0300	−1.0503

(a) Use the table to estimate the partial derivatives $g_x(1, 1)$ and $g_y(1, 1)$.

(b) It's true that $g_x(0, 0) = 0$ and $g_y(0, 0) = 1$. How do the table entries reflect these facts?

(c) Consider the linear function $L(x, y) = 0 + 0x + 1y = y$. Show that $L_x(0, 0) = g_x(0, 0) = 0$, $L_y(0, 0) = g_y(0, 0) = 1$, and $L(0, 0) = g(0, 0) = 0$.

(d) Fill in the following table of values for the function L from part (c).

Values of $L(x, y)$					
$y \backslash x$	−0.02	−0.01	0.00	0.01	0.02
0.02					
0.01					
0.00					
−0.01					
−0.02					

Compare your results with the tabulated values of g. [The results show how L linearly approximates g near $(0, 0)$.]

(e) Find the linear function $M(x, y)$ such that (i) $M(1, 1) = g(1, 1)$, (ii) $M_x(1, 1) = g_x(1, 1)$, and (iii) $M_y(1, 1) = g_y(1, 1)$. [HINT: One approach is to write $M(x, y) = a + b(x-1) + c(y-1)$ and then use the conditions to find values for a, b, and c.]

7. Let $f(x, y) = \sin y + 2$. (The formula is independent of x.)

(a) Plot f; use the domain $-5 \leq x \leq 5$, $-5 \leq y \leq 5$. How does the shape of the graph reflect the fact that f is independent of x? (In Section 1.1 we called such graphs **cylinders**.)

(b) Find $f_x(x, y)$ and $f_y(x, y)$. How do the answers reflect the fact that f is independent of x?

(c) Find the linear approximation function L for f at the point $(0, 0)$. How does its form reflect the fact that f is independent of x?

8. This exercise is about the situation described in Example 3 (page 403). Use the function f and the contour map given there.

 (a) Use the contour map to estimate the partial derivatives $f_x(1.5, 1.5)$ and $f_y(1.5, 1.5)$.

 (b) Use the formula $f(x, y) = x^2 - 3xy + 6$ to find $f_x(1.5, 1.5)$ and $f_y(1.5, 1.5)$ exactly.

 (c) Use results of part (b) to find the linear approximation $L(x, y)$ to $f(x, y)$ at $(1, 5, 1.5)$.

9. Let $f(x, y) = \sin(x)$.

 (a) Draw a contour map of f in the rectangle $-\pi \le x \le \pi$, $-2 \le y \le 2$. Show the level curves that correspond to $z = \pm 1$, $z = \pm 0.75$, $z = \pm 0.5$, $z = \pm 0.25$, and $z = 0$.

 (b) Use the level curve diagram to estimate $f_x(0, 0)$ and $f_y(0, 0)$.

 (c) Use the level curve diagram to estimate $f_x(\pi/2, 0)$ and $f_y(\pi/2, 0)$.

 (d) The formula shows that $f_y(x, y) = 0$ for all (x, y). How does the contour map reflect this fact?

 (e) The formula shows that $f_x(x, y)$ is independent of y. How does the contour map of f reflect this fact?

10. Let $f(x, y) = \cos(y)$.

 (a) Draw a contour map of f in the rectangle $-\pi \le x \le \pi$, $-2 \le y \le 2$. Show the level curves that correspond to $z = \pm 1$, $z = \pm 0.75$, $z = \pm 0.5$, $z = \pm 0.25$, and $z = 0$.

 (b) Use the level curve diagram to estimate $f_x(0, 0)$ and $f_y(0, 0)$.

 (c) Use the level curve diagram to estimate $f_x(\pi/2, 0)$ and $f_y(\pi/2, 0)$.

 (d) The formula for f shows that $f_x(x, y) = 0$ for all (x, y). How does the contour map of f reflect this fact?

 (e) The formula for f shows that $f_y(x, y)$ is independent of x. How does the contour map of f reflect this fact?

11. Let $f(x, y) = 2x - 3y$.

 (a) Draw a contour map of f in the rectangle $[-3, 3] \times [-3, 3]$. Show the level curves that correspond to $z = -5$, $z = -4$, $z = -3, \ldots, z = 4$, and $z = 5$.

 (b) Use your contour map (not the formula) to find $f_x(0, 0)$ and $f_y(0, 0)$.

 (c) The formula for f implies that both f_x and f_y are constant functions. How does the contour map of f reflect this fact?

 (d) The formula for f implies that for any (x, y) $f_x(x, y) = 2$ and $f_y(x, y) = -3$. How does the contour map reflect the fact that $f_x(x, y)$ is positive but $f_y(x, y)$ is negative?

12. Let $f(x, y) = 2y - x$.

    (a) Draw a contour map of f in the rectangle $[-3, 3] \times [-3, 3]$. Show the level curves that correspond to $z = -5$, $z = -4$, $z = -3, \ldots, z = 4$, and $z = 5$.

    (b) Use your contour map (not the formula) to find $f_x(0, 0)$ and $f_y(0, 0)$.

    (c) The formula for f implies that $f_x(x, y) = -1$ and $f_y(x, y) = 2$ for all (x, y). How does the contour map reflect these facts? In particular, how does the contour map show that $f_x(x, y)$ is negative but $f_y(x, y)$ is positive?

13. Let $f(x, y) = xy$. This exercise explores ideas like those of Example 4 (page 405).

    (a) Find the linear approximation function L to f at $(x_0, y_0) = (2, 1)$.

    (b) (Do this part by hand.) On one set of xy-axes, draw the level curves $L(x, y) = k$ for $k = 1, 2, 3, 4, 5$. On another set of axes, draw the level curves $f(x, y) = k$ for $k = 1, 2, 3, 4, 5$. (In each case, draw the curves into the square $[0, 3] \times [0, 3]$.)

    (c) How do the contour maps in part (b) reflect the fact that L is the linear approximation to f at the point $(2, 1)$? Explain briefly in words.

    (d) Use technology to plot contour maps of f and L in the window $1.8 \le x \le 2.2$, $0.8 \le y \le 1.2$. (This small window is centered at $(2, 1)$.) Explain what you see.

14. Repeat Exercise 13 using the function $f(x, y) = x^2 - y^2$.

15. Let $f(x, y) = x^2 + y^2$. (See the contour map in Section 1.1.)

    (a) Use the contour map of f to estimate the partial derivatives $f_x(1, 2)$ and $f_y(1, 2)$.

    (b) Check your answers to part (a) by symbolic differentiation.

    (c) Use your answers from part (a) to find the linear approximation $L(x, y)$ to $f(x, y)$ at $(1, 2)$.

    (d) On one set of axes, plot the level curves $L(x, y) = k$ and $f(x, y) = k$ for $k = 3, 4, 5, 6, 7$. (Use the window $[0, 3] \times [0, 3]$.) What's special about the point $(1, 2)$?

16. Let $f(x, y) = \sin(x) + 2y + xy$.

    (a) Find the partial derivatives $f_x(x, y)$ and $f_y(x, y)$; then evaluate $f_x(0, 0)$ and $f_y(0, 0)$.

    (b) Find a linear function $L(x, y) = a + bx + cy$ such that $L_x(0, 0) = f_x(0, 0)$, $L_y(0, 0) = f_y(0, 0)$, and $L(0, 0) = f(0, 0)$.

    (c) Complete the following table (report answers to 4 decimals).

(x, y)	$(0, 0)$	$(0.01, 0.01)$	$(0.1, 0.1)$	$(1, 1)$
$f(x, y)$				
$L(x, y)$				

How do the answers reflect the fact that L approximates f closely near $(0, 0)$?

    (d) Use technology to draw contour plots of both f and L on the rectangle $-1 \le x \le 1$, $-1 \le y \le 1$. Label several contours on each. How do the pictures reflect the fact that L approximates f closely near $(0, 0)$?

17. For each function f, find the linear function L that linearly approximates f at the given point (x_0, y_0). (If possible, check your answers graphically by plotting both f and L near (x_0, y_0).)

    (a) $f(x, y) = x^2 + y^2$; $(x_0, y_0) = (2, 1)$.

    (b) $f(x, y) = x^2 + y^2$; $(x_0, y_0) = (0, 0)$.

    (c) $f(x, y) = \sin(x) + \sin(y)$; $(x_0, y_0) = (0, 0)$.

    (d) $f(x, y) = \sin(x)\sin(y)$; $(x_0, y_0) = (0, 0)$.

18. Let $f(x, y)$ be a differentiable function of two variables, let (x_0, y_0) any point in its domain, and let $L(x, y)$ be the linear approximation to f at (x_0, y_0). Show that if f is independent of one of the variables—say, x—then so is L.

19. Suppose we know that for a certain function f, $f(3, 4) = 25$, $f_x(3, 4) = 6$, $f_y(3, 4) = 8$, and $f(4, 5) = 41$.

    (a) Find a linear function $L(x, y)$ that approximates f as well as possible near $(3, 4)$.

    (b) Use L to estimate $f(2.9, 3.9)$, $f(3.1, 4.1)$, and $f(4, 5)$.

    (c) Could f itself be a linear function? Why or why not?

20. Suppose we know that for a certain function g, $g(3, 4) = 5$, $g_x(3, 4) = 3/5$, $g_y(3, 4) = 4/5$, and $g(4, 5) = \sqrt{41}$.

    (a) Find a linear function $L(x, y)$ that approximates g as well as possible near $(3, 4)$.

    (b) Use L to estimate $g(2.9, 4.1)$ and $g(4, 5)$.

    (c) Could g be a linear function? Why or why not?

21. Let $f(x, y) = |y| \cos x$. This exercise explores the fact that the partial derivatives of a function may or may not exist at a given point.

    (a) Use technology to plot $z = f(x, y)$ over the rectangle $[-5, 5] \times [-5, 5]$. The graph suggests that there may be trouble with partial derivatives where $y = 0$, i.e., along the x-axis. How does the graph suggest this? Which partial derivative (f_x or f_y) seems to be in trouble?

    (b) Use the definition to show that $f_y(0, 0)$ does not exist. In other words, explain why the limit

$$\lim_{h \to 0} \frac{f(h, 0) - f(0, 0)}{h}$$

    does not exist.

    (c) Show that $f_x(0, 0)$ *does* exist; find its value. How does the result appear on the graph?

    (d) Use the limit definition to show that $f_y(0, \pi/2)$ *does* exist; find its value.

    (e) How does the graph reflect the result of part (d)? (You may need to do some experimenting with the graph to answer this.)

(f) Find a function $g(x, y)$ for which $g_y(0, 0)$ exists but $g_x(0, 0)$ does not. Use technology to plot its graph.

22. Let $f(x, y) = |x|y$.

    (a) By experimenting with graphs (use technology!), try to guess where $f_x(x, y)$ and $f_y(x, y)$ do exist and where they don't. (No proofs are needed.)

    (b) Find $f_x(0, 0)$.

    (c) Explain why $f_x(0, 1)$ does not exist.

14.4   Optimization and partial derivatives: a first look

1. Let $g(x, y) = xy$. (Its contour map is shown in Example 1.) To an ant walking along the surface $z = g(x, y)$ from lower left to upper right, the origin seems to be a low spot; another ant walking from upper left to lower right would experience the origin as a high spot.

   (a) An ant walks along the surface from $(0, -1)$ to $(0, 1)$. How does the ant's altitude change along the way?

   (b) Another ant walks along the surface from $(0.5, -1)$ to $(0.5, 1)$. How does the ant's altitude change along the way? Where is the ant highest? How high is the ant there?

2. See the contour map of $f(x, y) = \cos(x) \sin(y)$ on page 417.

   (a) The surface $z = f(x, y)$ resembles an egg carton. Where do the eggs go?

   (b) From the picture alone, estimate the coordinates of a local minimum point, a local maximum point, and a saddle point.

   (c) Use the formula $f(x, y) = \cos(x) \sin(y)$ to find (exactly) all the stationary points of f in the rectangle $R = [-3, 3] \times [-3, 3]$.

   (d) Find the maximum and minimum values of f in the rectangle $R = [-3, 3] \times [-3, 3]$.

3. Consider the function $f(x, y) = x(x - 2) \sin(y)$. Here's a contour map:

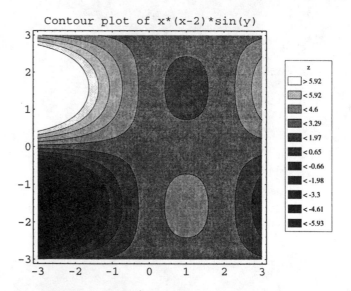

   (a) The function f has two stationary points along the line $x = 1$. Use the picture to estimate their coordinates. What type is each one?

   (b) There are four stationary points inside the rectangle $R = [-3, 3] \times [-3, 3]$. Use the formula for f to find all four.

   (c) The contour plot shows that f assumes its maximum and minimum values on $R = [-3, 3] \times [-3, 3]$ somewhere along the left boundary, i.e., where $x = -3$. Find these maximum and minimum values.   [HINT: If $x = -3$, then $f(x, y) = f(-3, y) = 15 \sin(y)$. This is a function of one variable, defined for $-3 \le y \le 3$.]

4. For each of the following functions, use the formula to find all stationary points. Then use technology (e.g., a properly chosen contour plot or surface plot) to decide what type of stationary point each one is.

   (a) $f(x, y) = -x^2 - y^2$
   (b) $f(x, y) = x^2 - y^2$
   (c) $f(x, y) = 3x^2 + 2y^2$
   (d) $f(x, y) = xy - y - 2x + 2$

5. Consider the linear function $L(x, y) = 1 + 2x + 3y$. Does L have any stationary points? If so, what type are they? If not, why not?

6. Consider the linear function $L(x, y) = a + bx + cy$, where a, b, and c are any constants.

   (a) The graph of L is a plane. Which planes have stationary points? For these planes, where are the stationary points?

   (b) Under what conditions on a, b, and c will L have stationary points? In this case, where are the stationary points? Reconcile your answers with those in part (a).

7. Let $f(x, y) = x^2$. The graph of f is a cylinder, unrestricted in the y-direction.

   (a) Use technology to plot the surface $z = f(x, y)$. Where in the xy-plane are the stationary points? What type are they?   [HINT: There's a whole line of stationary points.]

   (b) Use partial derivatives of f to find all the stationary points. Reconcile your answer with part (a).

8. Give an example as described in each part.   [HINTS: (1) See Exercise 7 for ideas. (2) Check your answers by plotting.]

   (a) A function $g(x, y)$ for which every point on the x-axis is a local minimum point

   (b) A function $h(x, y)$ for which every point on the line $x = 1$ is a local maximum point

   (c) A non-constant function $k(x, y)$ which has a local minimum at $(3, 4)$

14.5   Multiple integrals and approximating sums

Important Notes.   *Maple* (or a similar utility) is essential for some of the exercises. The appropriate commands can be loaded into *Maple* from the Maple worksheet `riemannsums.ms`. Begin by working through that worksheet.

In all cases, approximating sums are evaluated at the *midpoints* of subintervals or subregions.

1. Calculate by hand (without technology) the midpoint sum with 4 subdivisions for the $\int_0^1 x^2 \, dx$. Then check that *Maple* agrees. Finally, compare *Maple*'s answer for 100 subdivisions.

2. Calculate by hand (without technology) the double midpoint sum with $n = 3$ (i.e., 9 subdivisions in all) for the integral $\iint_R \sin(x) \sin(y) \, dA$ over the rectangle $R = [0, 1] \times [0, 1]$. Check your answer using *Maple*. Finally, compare *Maple*'s answer for $n = 10$, i.e., 100 subdivisions in all.

3. Calculate by hand (without technology) the triple midpoint sum with $n = 2$ (i.e., 8 subdivisions in all) for the triple integral $\iiint_R xyz \, dV$ over the cube $R = [0, 4] \times [0, 4] \times [0, 4]$. Check your answer using *Maple*. Finally, compare *Maple*'s answer for $n = 4$, i.e., $4^3 = 64$ subdivisions in all.

4. Let $f(x, y) = x + y$, let $R = [0, 4] \times [0, 4]$, and let $I = \iint_R f(x, y) \, dA$. Here is a contour plot of f:

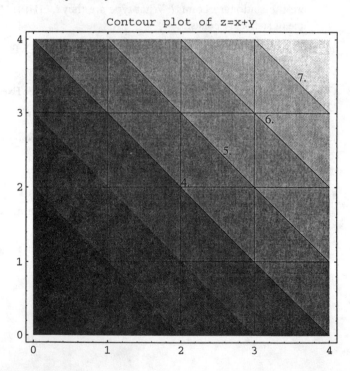

(a) Use the contour plot to evaluate a double midpoint sum for I with $n = 4$ (16 subdivisions in all).

(b) Use *Maple* to check your answer from part (a).

(c) Your answer in part (a) is, in fact, the exact value of the integral I. How does the symmetry of the contour map show this?

5. Let $f(x, y) = x^2 + y^2$, let $R = [0, 4] \times [0, 4]$, and let $I = \iint_R f(x, y)\, dA$. Here is a contour plot of f:

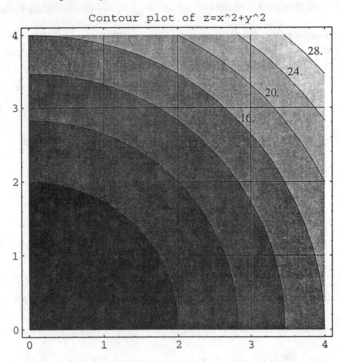

Contour plot of z=x^2+y^2

(a) Use the contour plot to evaluate a double midpoint sum for I with $n = 4$ (16 subdivisions in all).

(b) Use *Maple* to check your answer from part (a).

(c) Would you expect your answer from part (a) to overestimate or underestimate the true value of I? How can you tell?

14.6   Calculating integrals by iteration

Note.   *Maple* (or a similar utility) will be helpful in several of these exercises.

1. Use iteration to calculate each of the following integrals by hand (without technology). Then check your answers symbolically, using *Maple*. Finally, plot a 3d-surface over an appropriate domain to see that your answer is reasonable. [NOTE: Most of these integrals appear among the exercises in Section 14.5.]

   (a) $\iint_R \sin(x)\sin(y)\,dA$; $R = [0,1] \times [0,1]$

   (b) $\iint_R \sin(x+y)\,dA$; $R = [0,1] \times [0,1]$

   (c) $\iint_R (x^2+y^2)\,dA$; $R = [0,4] \times [0,4]$

   (d) $\iiint_R x\,dV$; $V = [0,1] \times [0,2] \times [0,3]$

   (e) $\iiint_R y\,dV$; $V = [0,1] \times [0,2] \times [0,3]$

2. Use iteration to calculate each of the following nonrectangular integrals by hand (i.e., without technology). In each case, the inner integral should be in y and the outer integral in x. Check your answers symbolically, using *Maple*.

   (a) $\iint_R (x+y)\,dA$; R the region bounded by the curves $y = x$ and $y = x^2$

   (b) $\iint_R x\,dA$; R the region bounded by the curves $y = x^2$ and $y = \sqrt{x}$

   (c) $\iint_R 1\,dA$; R the first quadrant part of the circle $x^2 + y^2 \le 1$

3. Redo Exercise 2, but integrate first in x and then in y.

4. Consider the integral $I = \iint_R (x+y)\,dA$, where R is the region bounded by the curves $y = x^2$ and $y = 1$.

   (a) Calculate I by integrating first in y and then in x.

   (b) Calculate I by integrating first in x and then in y.

5. Let $f(x,y) = x$, and let R be the plane region bounded by the curves $y = e^x$, $y = 0$, $x = 0$, and $x = 1$.

   (a) Calculate $I = \iint_R f(x,y)\,dA$ by integrating first in y and then in x.

   (b) Calculate $I = \iint_R f(x,y)\,dA$ by integrating first in x and then in y. [HINT: First split the region R into two simpler pieces; each simpler piece should be bounded on the left by one curve and on the right by another.]

6. Let $y = f(x)$ be a function, with $f(x) \ge 0$ if $a \le x \le b$; let R be the plane region bounded by the curves $y = f(x)$, $y = 0$, $x = a$, and $x = b$.

   (a) What does *single-variable* calculus say about the area of R?

   (b) According to Section 14.5, the double integral $I = \iint_R 1\,dA$ gives the area of R. Use an iterated integral to reconcile this formula with the one in part (a).

7. Let $x = g(y)$ be a function with $g(y) \geq 0$ if $c \leq y \leq d$; let R be the plane region bounded by the curves $x = g(y)$, $x = 0$, $y = c$, and $y = d$.

   (a) What does *single-variable* calculus say about the area of R?

   (b) According to Section 14.5, the double integral $I = \iint_R 1 \, dA$ gives the area of R. Use an iterated integral to reconcile this formula with the one in part (a).

14.7 Double integrals in polar coordinates

1. Let R be the polar rectangle defined by $a \le r \le b$ and $\alpha \le \theta \le \beta$.

   (a) Show that the area of R is $\dfrac{a+b}{2}(b-a)(\beta-\alpha)$. [Hint: Appendix B discusses the area of a circular sector, or "wedge."]

   (b) Use part (a) to show that a polar rectangle with dimensions Δr and $\Delta \theta$ and inner radius r has area $\dfrac{r+r+\Delta r}{2}\,\Delta r\,\Delta\theta$.

2. Let $f(x,y)=y$, let R be the upper half of the region inside the unit circle $x^2+y^2=1$, and let $I=\iint_R f\,dA$.

   (a) Calculate I as an iterated integral in rectangular coordinates, with the inner integral in y.

   (b) Calculate I as an iterated integral in rectangular coordinates, with the inner integral in x.

   (c) Calculate I as an iterated integral in polar coordinates.

3. In this section, we used polar coordinates to calculate that
$$I_2 = \iint_{R_2} \sqrt{x^2+y^2}\,dA = 2\pi/3,$$
where R_2 is the region inside the unit circle $r=1$. (See the picture on page 443.) Use formulas for the volumes of cones and cylinders to find the same answer by elementary means.

4. This exercise is about the integral I_1 of Example 1 (page 440).

   (a) Draw (use technology if necessary, but try without it) the solid whose volume is given by I_1.

   (b) Evaluate I_1 again but with the inner integral in x, not y.

5. Use a polar double integral in each of the following parts. (Draw each region first.)

   (a) Find the area of the region inside the cardioid $r=1+\sin\theta$.

   (b) Find the area of the region bounded by $y=x$, $y=0$, and $x=1$. Could you find the answer another way? [Hint: First write the boundary equations in polar form.]

   (c) Find the area of the region bounded by the circle of radius $1/2$, centered at $(0,1/2)$. Could you find the answer another way? [Hint: One approach is to first write a Cartesian equation for the circle, then change it to polar form.]

6. Use polar coordinates to calculate each of the following quantities.

   (a) Find $\iint_R \dfrac{1}{\sqrt{x^2+y^2}}\,dA$, where R is the region inside the cardioid $r=1+\sin\theta$ and above the x-axis.

   (b) Find the volume of the solid under the surface $z=1-x^2-y^2$ and above the xy-plane. [Hint: First decide where the surface hits the xy-plane.]

   (c) Find the volume of the conical solid under the surface $z=1-\sqrt{x^2+y^2}$ and above the xy-plane. [Hint: First decide where the surface hits the xy-plane.]

3.8   Inverse Trigonometric Functions and Their Derivatives

1. $\arcsin(1) = \pi/2$

3. $\arccos(-1) = \pi$

5. $\arctan(1) = \pi/4$

7. $\arcsin(0.8) \approx 0.93$

9. $\arctan(0.4) \approx 0.38$

11. (a) No. The gaps in the *table* reflect gaps in the functions' *domains*. For instance, $\arcsin(2)$ is *undefined* because there *is* no number x such that $\sin x = 2$.

    (b) For any $x \in [-1, 1]$, $\arcsin x + \arccos x = \pi/2 \approx 1.57$.

13. $\sin(\arccos x) = \sqrt{1 - x^2}$

15. $\sin(\arctan x) = x/\sqrt{1 + x^2}$

17. $f'(x) = 2/(1 + 4x^2)$

19. $f'(x) = \frac{1}{2}\big((1 - x^2)\arcsin x\big)^{-1/2}$

21. $f'(x) = e^x/\sqrt{1 - e^{2x}}$

23. $f'(x) = \Big(x\big(1 + (\ln x)^2\big)\Big)^{-1}$

25. $f'(x) = 2/\sqrt{x^4 - 4x^2}$

27. $f'(x) = 2x\arctan(\sqrt{x}) + x^{3/2}/(2 + 2x)$

29. $F(x) = \arctan x$

31. $F(x) = \arctan(2x)$

33. $F(x) = \arcsin(x/3)$

35. $F(x) = \arctan(e^x)$

37. $F(x) = (\arctan x)^2$

39. $F(x) = e^{\arcsin x}$

41. $F(x) = \ln(\arctan x)$

43. $x \approx 1.107$, $x \approx 4.249$, $x \approx -2.034$, and $x \approx -5.176$.

45. (a) The range of the arcsine function is $[-\pi/2, \pi/2]$. Since 5 is not in this interval, no input to the arcsine function will yield 5.

    (b) The equation $\arcsin(\sin x) = x$ holds when $-\pi/2 \le x \le \pi/2$.

    (c) This is a straightforward application of the chain rule and the identity $\sqrt{1 - \sin^2 x} = |\cos x|$.

    (d) *In* the interval $-\pi/2 \le x \le \pi/2$, the graph shown is the line $y = x$. For other values of x, the graph can be thought of as the result of "folding" the line $y = x$ back and forth, to stay within the vertical range (the interval $[-\pi/2, \pi/2]$) of the inverse sine function.

47. (b) $\lim_{x \to \infty} f(x) = \pi/2$; $\lim_{x \to -\infty} f(x) = -\pi/2$; $\lim_{x \to \infty} f'(x) = \lim_{x \to -\infty} f'(x) = 0$. These results imply the lines $y = -\pi/2$ and $y = \pi/2$ are horizontal asymptotes of the arctangent function.

  (c) f is an odd function.

  (d) f is increasing on $(-\infty, \infty)$ because f' is positive on this interval.

  (e) f is concave up on $(-\infty, 0)$ and concave down on $(0, \infty)$. f has an inflection point at $x = 0$.

49. (b) $\lim_{x \to 1^-} f(x) = 0$; $\lim_{x \to -1^+} f(x) = \pi$; $\lim_{x \to 1^-} f'(x) = \lim_{x \to -1^+} f'(x) = -\infty$. These results imply that the lines $x = -1$ and $x = 1$ are vertical asymptotes of the arccosine function.

  (c) f is neither even nor odd.

  (d) f is decreasing on $(-\infty, \infty)$ because f' is negative on this interval.

  (e) f is concave up on $(-1, 0)$ and concave down on $(0, 1)$. f has an inflection point at $x = 0$.

  (f) The graph of f' is increasing on $(-1, 0)$ and decreasing on $(0, 1)$.

51. Let $f(x) = \arctan x - x/(1 + x^2)$ and $g(x) = x - \arctan x$. Now,

$$f'(x) = \frac{1}{1 + x^2} - \frac{1 + x^2 - x(2x)}{(1 + x^2)^2} = \frac{1 + x^2 - (1 - x^2)}{(1 + x^2)^2} = \frac{2x^2}{(1 + x^2)^2} > 0$$

for all x, and

$$g'(x) = 1 - \frac{1}{1 + x^2} = \frac{1 + x^2}{1 + x^2} - \frac{1}{1 + x^2} = \frac{x^2}{1 + x^2} > 0$$

for all x. Since $f'(x) > 0$ and $g'(x) > 0$ for all x, and since $f(0) = g(0) = 0$, $f(x) > 0$ and $g(x) > 0$ for all $x \geq 0$.

53. Differentiate both sides of the identity given and verify that the derivatives are equal. This establishes that the two sides differ from each other by a constant. Evaluating both sides for a particular value of x (e.g., $x = 0$) shows that this constant is zero.

55. Differentiate both sides of the identity given and verify that the derivatives are equal. This establishes that the two sides differ from each other by a constant. Evaluating both sides for a particular value of x (e.g., $x = 0$) shows that this constant is zero.

57. Differentiate both sides of the identity given and verify that the derivatives are equal. This establishes that the two sides differ from each other by a constant. Evaluating both sides for a particular value of x (e.g., $x = 0$) shows that this constant is zero.

59. (a) $f'(x) = 1/(1 + x^2) = (\arctan x)' \implies f(x) = C + \arctan x$. Since $f(0) = \pi/4$, $f(x) = \pi/4 + \arctan x$ when $x < 1$.

  (b) $\lim_{x \to 1^-} f(x) = \pi/2$

  (c) When $x > 1$, $x = 1/y$ where $0 < y < 1$. Therefore, $f(x) = f(1/y) = \arctan\left(\frac{1 + 1/y}{1 - 1/y}\right) = \arctan\left(\frac{1 + y}{y - 1}\right) = -\arctan\left(\frac{1 + y}{1 - y}\right) = -f(y)$. Thus, $\lim_{x \to 1^+} f(x) = \lim_{y \to 1^-} -f(y) = -\pi/2$.

61. (a) Let $x = \arctan u$ and $y = \arctan v$. Then, $\tan(\arctan u + \arctan v) = (u + v)/(1 - uv) \implies \arctan u + \arctan v = \arctan\left(\frac{u + v}{1 - uv}\right)$.

  (b) The identity in part (a) is valid when $-\pi/2 < \arctan x + \arctan y < \pi/2$.

4.1   Differential Equations and Their Solutions

1. $y' = t$

3. $y' = -e^t \neq -y$

5. $y' = -42e^{-t} = -y$

7. $y' = e^t + t \neq y + t$

9. $y' = t \exp(t^2/2) = ty$

11. (a) The coffee reaches $100°$ F at $t = 2\ln(35/125)/\ln(95/125) \approx 9.28$ minutes.

    (b) The formula for $z(t)$ has the form $z(t) = 65 + Ae^{kt}$. The values $z(0) = 190$ and $z(2) = 180$ imply that $A = 125$ and $k = \ln\left(\sqrt{115/125}\right) \approx -0.04169$. Thus, $z(t) = 65 + 125e^{\ln\sqrt{115/125}\,t} \approx 65 + 125e^{-0.04169t}$.

    (c) Solving $z(t) = 100$ for t gives $t = 2\ln(7/25)/\ln(23/25) \approx 30.53$ minutes.

    (d)

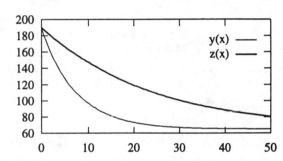

13. $P'(t) = -C\left(1 + de^{-kCt}\right)^{-2} \cdot \left(-kCde^{-kCt}\right) = \dfrac{kC^2 de^{-kCt}}{\left(1 + de^{-kCt}\right)^2}.$

    $C - P = Cde^{-kCt}/\left(1 + de^{-kCt}\right)$ so $kP(C - P) = P'$.

15. (a) $v'(t) = 1$ means that the object accelerates upward at 1 meter per second per second.

    (b) $h'(t) = -3$ means that the object falls at a constant rate of 3 meters per second.

    (c) $v'(t) = -0.01v^2(t)$, means that the acceleration is proportional, but opposite in direction, to the square of the velocity.

17. (a) $v' = -5$ (or $v'(t) = -5$).

    (b) $p' = 15$ (or $p'(t) = 15$).

    (c) $v' = kv$; k is negative because friction works in the direction *opposite* to the velocity.

    (d) $p' = k\sqrt{p}$.

19. The rate at which the influenza spreads is described by the differential equation $P' = kP(3000 - P)$ where k is a constant.

4.2 More Differential Equations: Modeling Growth

1. $y(x) = 100e^{0.1x}$

3. $y(x) = 2^x$

5. Mimicking the solution procedure in the text, $k = 1/90$ and $d = 9$. Thus, $P(t) = 10/\left(1 + 9e^{-t/9}\right)$.

   (a) $P(10) = 10/(1 + 9e^{-10/9}) \approx 2.52$ thousand flies; $P(100) = 10/(1 + 9e^{-100/9}) \approx 9.999$ thousand flies.
   (b) $P(t) = 5000$ when $t = 9\ln(9) \approx 19.78$ days.
       $P(t) = 9000$ when $t = 9\ln(81) \approx 39.55$ days.
       $P(t) = 9900$ when $t = 9\ln(891) \approx 61.13$ days.
   (c) The population is growing fastest when $P(t) = 5000$. This occurs when $t = 9\ln(9) \approx 19.78$ days.

7. Let $B(t)$ be the weight of The Blob at time t hours after noon on Wednesday. Since The Blob grows at a rate proportional to its size, its growth is described by the differential equation $B' = kB$ and, therefore, $B(t) = Ae^{kt}$ for some numbers A and k. Since $B(0) = 1$ and $B(4) = 4$, $A = 1$ and $k = \frac{1}{4}\ln 4$. Thus, The Blob will weigh 3×10^{15} at time $t = 4\ln\left(3 \times 10^{15}\right)/\ln 4 \approx 102.83$ hours (or, equivalently, in about 4.28 days).

9. (a) Since the bottle becomes full at $t = 24$ hours and the volume of the culture doubles every hour, the bottle must be half-full at time $t = 23$ hours.

    (b) Let $V(t)$ be the volume of the culture at time t. Then, since the bottle is full at time $t = 24$, the bottle is less than 1% full when $V(t) < 0.01V(24)$ or $V(t)/V(24) < 0.01$.

        The volume of the culture grows at a rate proportional to the amount present, so by Theorem 1, page 290, $V(t) = V(0)e^{kt}$. Since $V(1) = 2V(0)$, $k = \ln 2$ so $V(t) = V(0)2^t$. It follows that the bottle is less than 1% full when $2^{t-24} < 0.01$; that is, when $t < 24 + \log_2(0.01) \approx 17.356$ hours. Thus, the bottle is less than 1% full approximately 72% of the time.

11. Let $R(t)$ denote the amount of oil remaining in the well at time t. Then $R'(t) = kR(t)$ is the rate at which oil is being pumped from the well at time t and $R(t) = R(0)e^{kt}$. The conditions $R(0) = 10^6$ and $R(6) = 5 \times 10^5$ imply that $k = -\frac{1}{6}\ln 2$.

    (a) Since $R' = kR$, $R'(6) = \left(-\frac{1}{2}\ln 2\right)\left(6 \times 10^5\right) = -10^5 \ln 2 \approx -69315$ barrels per year. Thus, oil is being pumped out at a rate of approximately 69,315 barrels per year.
    (b) $R(t) = 5 \times 10^4$ when $t = -6\ln 0.05/\ln 2 \approx 25.932$ years.

13. (a) At time $t = 0$, the concentration of salt in the mixture is 0.1 lbs/gal. No salt is added to the mixture, so the amount of salt leaving the tank is proportional to the concentration of salt in the tank. Thus, $S(t)$, the amount of salt in the tank at time t, is the solution of the differential equation $S'(t) = -\frac{5}{100}S(t)$ where t is measured in minutes.

    (b) Since $S(0) = 10$, the solution of the differential equation from part (a) is $S(t) = S(0)e^{-0.05t} = 10e^{-0.05t}$.
    (c) After one hour, the amount of salt left in the tank will be $S(60) = 10e^{-0.05 \cdot 60} = 10e^{-3} \approx 0.49787$ pounds.

4.3   Linear and Quadratic Approximation; Taylor Polynomials

1. If $f(x) = \dfrac{1}{1-x}$, $n = 3$, and $x_0 = 0$, then $P_3(x) = 1 + x + x^2 + x^3$.

3. If $f(x) = \ln x$, $n = 3$, $x_0 = 1$, then $P_3(x) = x - 1 - \dfrac{(x-1)^2}{2} + \dfrac{(x-1)^3}{3}$.

5. If $f(x) = \sqrt{x}$, $n = 3$, $x_0 = 4$, then $P_3(x) = 1 + \dfrac{x}{4} - \dfrac{(x-4)^2}{64} + \dfrac{(x-4)^3}{512}$.

7. $f(x) = \cos x$ has $\ell(x) = 1$; $[-0.1415, 0.1415]$
   $q(x) = 1 - x^2/2$; $[-0.7028, 0.7028]$

9. $f(x) = e^x$ has $\ell(x) = 1 + x$, $q(x) = 1 + x + x^2/2$.

11. $f(x) = \arcsin x$ has $\ell(x) = x$, $q(x) = x$.

13. (b) $q(x) = (a - 64b + 4096c) + (b - 128c)x + cx^2$.

    (c) $q'(x) = (b - 128c) + 2cx$. Notice that this agrees with what one gets by differentiating the expression in the previous part.

15. (a) Easy calculations show that $P_6(x) = 1 - \dfrac{x^2}{2} + \dfrac{x^4}{24} - \dfrac{x^6}{720}$. The lower-order polynomials can now be read off.

    (b) Both g and all the Maclaurin polynomials are even. The graphs "reflect" this by being symmetric about the y-axis.

    (c) The derivatives $P_1'(x)$, $P_3'(x)$, $P_5'(x)$, $P_7'(x)$ of the *sine* polynomials are, respectively, the Maclaurin polynomials P_0, P_2, P_4, and P_6 for $g(x) = \cos x$. This is no great surprise, since $f'(x) = g(x)$.

17. (a) We know that $g(5) = 2$; the graph shows that $g'(5) = 1$. Thus the linear approximation to g at $x = 5$ is $\ell(x) = 2 + 1(x - 5)$; it gives the (crude) estimate $g(0) \approx \ell(0) = -3$.

    (b) We know that $g(5) = 2$; the graph shows that $g'(5) = 1$. The slope of the g' graph at $x = 5$ is about -2, so $g''(5) \approx -2$. Thus the quadratic approximation to g at $x = 5$ is $q(x) = 2 + (x - 5) - (x - 5)^2$; it gives the (crude) estimate $g(8) \approx q(8) = -4$.

19. (a) $\ell_p(t) = 25t$; $\ell_p(1) = 25$; $\ell_p(-1) = -25$
    (b) $q_p(t) = 25t + t^2$; $q_p(1) = 26$; $q_p(-1) = -24$
    (c) $\ell_v(t) = 25 + 2t$; $\ell_v(1) = 27$
    (d) $\pm 3/2$ meters per second

21. (a) $\ell(t) = 100$ meters; $\ell(1) = 100$ meters
    (b) $q(t) = 100 - 4.9t^2$ meters; $q(1) = 95.1$ meters

23. (a) $q(x) = (x - 1) - \frac{1}{2}(x - 1)^2$

    (b) $f'''(x) = 2/x^3$ so $16/27 \le f'''(x) \le 16$ if $1/2 \le x \le 3/2$. Therefore,

    $$|f(x) - q(x)| \le \frac{16}{6}(x - 1)^3 \le \frac{16}{6}\left(\frac{1}{2}\right)^3 = \frac{1}{3}.$$

    (c) The actual approximation error is less than that allowed by the error bound formula.

25. (a) If $f(x) = \sqrt{x}$, then $f'(x) = x^{-1/2}/2$; $f''(x) = -x^{-3/2}/4$. On the interval $[100, 103]$, $|f''(x)| \le 1/\left(4 \cdot 100^{3/2}\right) \approx 0.00025$. Hence $K = 0.00025$ works, and the theoretical error bound is $\frac{K}{2}(103 - 100)^2 = 0.000125 \cdot 9 = 0.001125$. The actual error in $\ell(x)$, computed above, was 0.00111.

(b) If $f(x) = x^{1/3}$, then $f'(x) = x^{-2/3}/3$; $f''(x) = -2/\left(9x^{5/3}\right)$. On the interval $[27, 29]$, $|f''(x)| \leq 2/\left(9 \cdot 27^{5/3}\right) \approx 0.001$. Thus $K = 0.001$ works, and the theoretical error bound is $\frac{K}{2}(29 - 27)^2 = 0.0005 \cdot 4 = 0.002$. The actual error in $\ell(x)$, computed above, was 0.00176, a bit less than the bound above permits.

(c) If $f(x) = \tan x$, then $f'(x) = \sec^2 x$; $f''(x) = 2\sec^2 x \tan x$. On the interval $[30\pi/180, 31\pi/180]$, $|f''(x)| \leq 1.64$ (plot f'' to see why). Thus $K = 1.64$ works, and the theoretical error bound is $\frac{K}{2}(\pi/180)^2 = \approx 0.00025$. The actual error in $\ell(x)$, computed above, was 0.00024, a little less than the bound above permits.

(d) If $f(x) = x^{10}$, then $f''(x) = 90x^8$. On the interval $[0.8, 1]$, $|f''(x)| \leq 90$. Thus $K = 90$ works, and the theoretical error bound is $\frac{K}{2}(0.8 - 1)^2 = 45(0.2)^2 = 1.8$. The actual error in $\ell(x)$, computed above, was 1.10737, less than the bound above permits.

27. (a) $f(x) = \ln x$, $f'(x) = 1/x$, and $f''(x) = -1/x^2$ so $f(1) = 0$, $f'(1) = 1$, and $f''(1) = -1$. Therefore, $\ell(x) = x - 1$, $q(x) = (x - 1) - (x - 1)^2/2$, $E_1(x) = \ln x - (x - 1) = \ln x - x + 1$, and $E_2(x) = \ln x - (x - 1) + (x - 1)^2/2 = \ln x + 3/2 - 2x + x^2/2$.

(c) $|E_2(x)| \leq |E_1(x)|$ if $0 \leq x \leq 2$

(d) The graph of E_1 resembles a quadratic curve and the graph of E_2 resembles a cubic curve.

(e) No, they just look that way. If $x_0 = 1$, the first nonzero term in the Taylor series for $E_1(x)$ is of degree 2 so its graph appears quadratic when x is near one. Similarly, if $x_0 = 1$, the first nonzero term in the Taylor series for $E_2(x)$ is of degree 3 so its graph appears cubic when x is near one.

4.9   Parametric equations, parametric curves

1. The curve is the upper unit semi-circle plotted from $(-1, 0)$ to $(0, 1)$ to $(1, 0)$.

3. The curve is the right unit semi-circle plotted from $(0, -1)$ to $(1, 0)$ to $(0, 1)$.

5. The curve is the unit circle plotted clockwise from $(0, -1)$ to $(0, 1)$ to $(0, -1)$.

7. In each case the idea is to calculate $\sqrt{f'(t)^2 + g'(t)^2}$; if the result is constant, then the curve has constant speed. Among the given choices only the last—$x = \sin(\pi t)$, $y = \cos(\pi t)$—has constant speed.

9. (a) The spacing of bullets suggests that P moves quickly at $t = 3$, $t = 4$, $t = 9$, and $t = 10$, and slowly at $t = 0$, $t = 1$, $t = 6$, and $t = 7$.

   (b) The distance along the curve from $t = 2.5$ to $t = 3.5$ seems to be about 3 units. Thus P appears to travel about 3 units per second at $t = 3$.

   (c) Use the curve to estimate the speed of P at $t = 6$. The distance along the curve from $t = 5.5$ to $t = 6.5$ seems to be about 1 unit. Thus P appears to travel about 1 unit per second at $t = 6$.

11. (a) The result is the circle of radius 2, centered at $(2, 1)$.

   (b) Here's the calculation: Since $x = a + r \cos t$ and $y = b + r \sin t$,

   $$(x - a)^2 + (y - b)^2 = r^2 (\cos t)^2 + r^2 (\sin t)^2 = r^2.$$

   (c) Setting $x = 2 + \sqrt{13} \cos t$, $y = 3 + \sqrt{13} \sin t$, and $0 \le t \le 2\pi$, gives the circle of radius $\sqrt{13}$, centered at $(2, 3)$.

   (d) No proper "curve" results: for all t, (x, y) stays put at $(2, 3)$.

13. (a) The origin corresponds to $t = 0$; $P(0.1) \approx (0.48, 0.56)$; $P(\pi/2) = (1, 0)$. Thus P starts at the origin and starts off in a northeasterly direction.

   (b) Both x and y are 0 if and only if both $5t$ and $6t$ are integer multiples of π. This occurs only for $t = 0$, $t = \pi$, and $t = 2\pi$.

   (c) Using the t-interval $0 \le t \le 4\pi$ would produce exactly the same curve, but it would be traversed twice.

15. (a) The curve starts at $(at_0 + b, ct_0 + d)$ and ends at $(at_1 + b, ct_1 + d)$.

   (b) $y = \dfrac{c}{a}(x - b) + d$

   (c) $x = \dfrac{a}{c}(y - d) + b$

   (d) If $a = c = 0$, the parametric curve is just the point (b, d).

17. (a) The model would be more realistic if it took wind resistance into account. To do so, one would need some mathematical information about wind resistance.

   (b) Imitate the argument given for $f(t)$. Notice, too, that if $g(t) = 7 - 16t^2$, then $g'' = -32$, $g(0) = 7$, and $g'(0) = 0$, just as claimed.

   (c) By definition, $s(t) = \sqrt{f'(t)^2 + g'(t)^2} = \sqrt{150^2 + (-32t)^2} = \sqrt{22500 - 1024t}$. Plotting this function over the interval $0 \le t \le 0.661$ (when the ball hits the ground) gives almost a horizontal line—the velocity changes very little over the short time interval.

19. (a) If $x = f(t) = s_0 t$ and $y = g(t) = 7 - 16t^2$ it's easy to check directly that $f''(t) = 0$, $f'(0) = s_0$, $f(0) = 0$, $g''(t) = -32$, $g'(0) = 0$, and $g(0) = 7$. These are the necessary conditions.

   (b) The ball reaches home plate when $f(t) = s_0 t = 60.5$, i.e., at $t = 60.5/s_0$ seconds.

(c) The trajectory is parabolic for any $s_0 > 0$. (If $s_0 = 0$, the ball drops straight down.) This can be seen by eliminating t. Since $x = s_0 t$, $t = x/s_0$, so $y = 7 - 16t^2 = 7 - 16x^2/s_0^2$. This is the equation of a parabola in the xy-plane.

21. Now, $x = 200 \ln(3t/4 + 1)$.

    (a) $x(t) = 60.5$ at time $t = 4\left(e^{121/400} - 1\right)/3 \approx 0.47098$. Thus, the air-dragged ball takes about 0.0677 seconds longer to reach the plate.

    (b) $y(t) \approx 3.4508$ feet at the time when $x(t) = 60.5$

    (c) When $x = 60.5$, the ball's speed is approximately 111.87 ft/sec.

    (d) When $y = 0$, $x \approx 80.569$ feet.